MEDICAL HYPNOTHERAPY

Techniques, Scripts and Processes for Effective Hypnosis and Healing

RICHARD K. NONGARD

Medical Hypnotherapy:
   Techniques, Scripts and Processes for Effective Hypnosis and
   Healing

By Richard K. Nongard, LMFT, CCH

Text Copyright © 2012 by Richard K. Nongard

Edited by James Hazlerig - www.HarmonyHypnosis.net

ISBN-13: 978-1-300-29997-4

Printed in the United States of America
by PeachTree Professional Education, Inc.

PeachTree Professional Education, Inc.

   7107 S. Yale, Ste 370
   Tulsa, OK 74136
   (918) 236-6116
   www.SubliminalScience.com
   www.HypotherapyBoard.org

All hypnotherapists are not created equal.

**When you learn with Richard K. Nongard,
you are learning with one of the
top hypnotherapists in the country.**

Richard K. Nongard is an innovative leader in the fields of psychotherapy and hypnosis. Since beginning his career in the late 1980s, he has trained literally thousands of professionals, including psychologists, medical doctors, ministers, social workers, family therapists, hypnotherapists and professional counselors in ways to do a better job serving their clients.

Nongard has authored many hypnosis and psychotherapy textbooks, videos and professional educational materials, and is the creator of the SuccessFit® Weight Loss Trance-Formation program, and the QuitSuccess® Tobacco Cessation Treatment Program, used by hospitals and healthcare groups across the country.

He holds advanced degrees in both counseling and religion, and has trained in the USA, Canada and Europe. He is a former Disciples of Christ minister, and has worked as a substance abuse counselor, a marriage and family therapist, and as a consultant to dozens of criminal justice departments. He is also the President of the International Certification Board of Clinical Hypnotherapy.

About the ICBCH

The International Certification Board of Clinical Hypnotherapy is a worldwide organization with thousands of members that include professional hypnotists, psychotherapists, social workers and counselors. We offer hypnosis training and education through live hypnosis seminars, online classes and home study courses on subjects related to hypnosis, NLP (Neuro-linguistic Programming), medical meditation and professional life coaching.

Each year the ICBCH holds an annual convention and also accredits schools and organizations that provide hypnosis training. You can become a certified hypnotherapist through the ICBCH by visiting: www.HypnotherapyBoard.org.

Our internet forum can be found at:
www.ICBCHForum.com

You are invited to connect with us on Facebook by visiting:
www.ICBCHGroup.com

Index and Course Outline

Introduction: 1

Chapter One:
Introducing Medical Hypnotherapy...3

Chapter Two:
Basic Medical Hypnotherapy Techniques 15

Chapter Three:
Suggestive Therapy for Depression..43

Chapter Four:
Pain Control 53

Chapter Five:
Sample Suggestions 66

Chapter Six:
Running a Medical Hypnotherapy Practice 81

Appendix A:
The Nongard Assessment of Primary Representational Systems.......................89

Appendix B:
44 Ways to Market Your Hypnosis Practice 91

Bibliography and Reading List...95

Introduction:

This book is a useful collection of scripts, processes and ideas for helping clients in hypnosis and hypnotherapy who present with medical difficulty. The original content was developed as part of web based learning course offered by the ICBCH. This book is a great starting point for moving beyond the basics of hypnosis and into specific applications of technique and processes.

Although this book contains many in-depth ideas and applications of technique, it is certainly not an exhaustive resource for all of the applications of medical hypnotherapy. Readers are encouraged to practice the scripts and patter contained herein, and to move forward in learning, adding additional resources and techniques to that which is contained in this text.

Of course, the ideas here are presented for educational purposes. They are not intended to either cover every possible medical condition or every possible process. The reader is advised that this book is not being offered for the diagnosis, prevention, treatment, or cure of any specific condition. You, and your clients are advised to consult a licensed medical doctor before applying any of the concepts or techniques offered in this book.

You are invited to ask questions and participate in our discussion forum. You can find this forum at www.ICBCHForum.com and you can also find other resources for connecting with medical hypnotists on www.ICBCHGroup.com.

Chapter One:
Introducing Medical Hypnotherapy

The Role of Hypnotherapy in Medicine

There is something important every hypnotherapist needs to internalize in order to:

> (A) build a successful medical hypnotherapy practice, and
> (B) do a good job with clients.

That very important idea you must internalize is this:

Hypnosis is NOT a complementary or alternative medical practice.

Let me reiterate:

Hypnosis is NOT a complementary or alternative medical method.

Hypnosis is not an alternative or complementary idea that has some uses with some medical clients. Hypnotherapy is in fact a *first-line intervention.* With some medical conditions, it is in fact the preferred method of helping clients to achieve success, wellness, and contentment.

Sometimes people say, "But hypnosis can't cure this condition or that condition." Well, I've never found anything yet—whether it's medication, whether it's surgery, whether it's hypnotherapy, whether it's any other approach to helping people make change—that can fix all problems all of the time. What we do have, though, are mountains of research that actually indicate that hypnotherapy is an effective tool when used conjointly with other approaches to helping a client experience success and can also be used as the primary intervention to help some of our clients experience success.

Hypnosis is not a 'one-size-fits-all' treatment. I'm not going to teach you a universal suggestion that you can apply to every client with every medical condition and find that they're cured at the end of one session. If I could do that, I probably wouldn't be running a practice in Tulsa, Oklahoma; I'd be sitting on top of my high mountain paradise, surveying my kingdom.

But what I am going to teach you are specific strategies that, when used with a wide variety of clients with medical conditions, will promote healing, wellness, and recovery. Some of those interventions with some of our clients are going to result in quick response, quick recovery, and quick healing. For some of our clients, as they participate in other forms of healing and recovery, hypnosis is going to aid in speeding the process, assisting them in experiencing comfort and helping our clients to achieve their goals.

Research Supporting Medical Hypnotherapy

While this course is centered around teaching you useful skills, we're going to briefly address the research behind medical hypnotherapy techniques. Not only is it useful to know what is actually in the clinical medical journals, nursing journals, psychology journals, and peer-reviewed research, it's very effective to convince yourself, your clients, and your medical professional colleagues of the efficacy of medical hypnotherapy.

Have you ever wondered: Can hypnosis be used with cancer patients or with burn clients, or with arthritis, Irritable Bowel Syndrome, migraine headaches, pre-surgical preparation, or heart surgery?

For any medical condition, you can always find out the efficacy of hypnotherapy by going to my favorite webpage: *scholar.google.com.*

(Don't go to *www.google.com.* Instead, substitute "scholar" for "www": so it is *scholar.google.com.*)

This will take you to a search engine that only searches two things:

> peer-reviewed scientific journals
> academic textbooks

Google Scholar will not search self-help books or self-published books--only university-backed, researched textbooks and academic publications.

You can type in, "HIV hypnosis," and the search will reveal hundreds of research articles with information about immune system response in hypnotherapy. You can type in "cancer hypnosis" and find mountains of peer-reviewed research demonstrating how hypnosis and hypnotherapy can be useful to the cancer patients that are on your caseload. You can type in any medical condition with the word "hypnosis" or "hypnotherapy" and what you will find is, in many of these cases, mountains of research indicating the effectiveness of hypnosis.

My Own Experience with Medical Hypnotherapy

In addition to the mountains of research, I want to relate my own personal experience with medical hypnotherapy. Those readers who know me in real life may recall that I used to walk with a cane because the arthritis in my foot had degenerated some of the joints. So I had foot surgery done on those joints. I actually had a couple of fusions, which is certainly not a desirable outcome, but the amazing part happened a week after my foot surgery when I went to the doctor's office.

The nurse said, "So can I get you a refill on your pain medication?"

And I said, "No."

"What do you mean no?" she asked. She'd never had anyone turn down free drugs.

I explained, "Well, in order to get a refill, I would've had to fill the first prescription."

"You didn't fill your pain medicine prescription?"

"Well, no." I answered.

"Aren't you in pain?" she exclaimed.

So I thought about it for a minute: The doctor had just sliced my foot open, opened up the skin, sawed out the joints, thrown them in the trash can, put in metal rods, screwed them back together, and then sewed me shut. So after thinking about it, I said, "Wow. I suppose I am in a little pain."

Looking at me like I was a crazy person, the nurse asked, "Are you going to fill your pain medication?"

And I replied, "No. I'm going to use the techniques I teach my clients for pain control."

By the way, I tell that story to some people, and they say, "How come you didn't use hypnosis to make the arthritic joints go away?"

One of the important things to learn in this course is the responsible use of hypnosis. You have to understand the limits of hypnosis. Sure, there are many miraculous uses of hypnosis, and we'll cover them in this course, particularly as we deal with the subject of cancer and cancer prevention.

There are other conditions, though, that are just slowly degenerative, like the arthritis I experienced in my foot. While it's true I walked with a cane, the reason why I was walking with a cane rather than rolling around in a wheelchair is that I used hypnosis. Sometimes we need to recognize that what hypnosis is doing is substantially decreasing the severity of the client condition.

Hypnosis is not about all or nothing. It's not the case that either you don't have arthritis or you do have arthritis. It's not the case that you're either well or not well. Many medical conditions are on a continuum. If I hadn't used hypnotic techniques, I would have been in much worse shape than I was in by the time I finally made the decision for surgery.

But don't take my word for this—take the word of my doctor. About six weeks after the foot surgery, and he came into his office and put an X-ray of my foot up on the screen.

He said, "Do you see that?"

"See what?" I asked.

He repeated, "Do you see that?"

I said to him, "Well, it's my foot."

6

Pointing, he said, "Do you see that right there?"

"Well, no, there's nothing right there."

"That's the issue," he said flatly.

I asked, "What do you mean?"

And he said, "That's where your fusion is."

"Okay." I wasn't sure what he was getting at.

He said, "You don't see it, do you?"

Still puzzled, I said, "No."

"That's because it's fused!" he exclaimed. "It usually takes three to six months for a fusion, but it's only been six weeks since your surgery! What are you doing?"

I was using hypnotic techniques to promote healing. I was doing what I ask my clients to do. I was doing what I'll teach you to do in this course.

How Medical Hypnosis Works

Now, during this course, I'm going to refer to some of the research from Ernest Rossi. Ernest Rossi was a contemporary of Milton Erickson, and they wrote a number of books together. He is an impeccable psychological researcher and has published numerous books on hypnosis and healing. A few years back he wrote a book titled *The Psychobiology of Mind-Body Healing: New Concepts for Therapeutic Hypnosis.*

It's not a textbook for the course, but a lot of this course is based on Rossi's information and ideas. I definitely recommend reading it—this book should be in every hypnotist's library. Another important author in this field is of course Bruce Lipton. Any hypnotist should be familiar with his work.

One of the things that I'm *not* going to cover in this course is the mechanism by which hypnosis promotes healing. Rossi actually covers that in tremendous detail. Bruce Lipton's work answers that same question about how our mind or mental processes affect our body.

I'm going to instead give you a very simple explanation for how hypnosis works:

Within the Western world, we make a distinction between mind and body, as if they are two different things. I think that the Eastern idea of mind and body being inseparable is the reality. We've made an impossible dichotomy, saying these are mental exercises, and these are physical exercises; these are our mental treatments, these are our physical treatments. You can't really separate the two.

The reason why physical healing takes place is because of mental processes; the mind and the body are inseparable. They are, in fact, one. And they work together as one. So when we impact the mind, we are impacting the body.

A lot of clients will come in and actually sit in the chair in my office and they'll say, "I've heard hypnosis can be helpful with cancer, HIV, burns, Irritable Bowel Syndrome, but these are real medical conditions. I don't understand how this is going to work. Am I going to sit in the comfortable furniture, close my eyes, count to ten, open my eyes, and be healed? Is that how it works?" I have clients who say to me, "My pain is real; it really hurts. How is it that hypnosis is actually going to reduce my pain?" What the client is actually saying is they don't understand that mind and body are one. They see it like most people in Western society do, and that is as two separate things.

The easiest way to illustrate the unity of mind and body is with the easiest and most basic of all hypnotic convincers. It's my favorite convincer during a pre-talk with medical clients. Allow yourself to engage your imagination as I guide you through this:

Remember when you were a kid? And you used to go to the movie theatre and you would eat those Lemonhead candies, you know what I'm talking about? The little Lemonhead box that had smiley face lemon on it, popular movie theatre candy. Go ahead: hold out your hand just imagine that I put in your hand a couple of imaginary lemon drop candies, those candies from that movie theatre. Now imagine taking with your other hand those imaginary candies, and pop those imaginary candies in your mouth right now. Go ahead and do that actually. Put the imaginary candy in your mouth and when you do, taste that sugary exterior, that hard candy, that sweet delicious exterior. And as that sugar begins to melt, of course, it gives way to the sour, tart, lemony inside of that hard candy. And you can taste that tart, lemon-flavored hard candy as that sugary exterior completely melts. Swish it around your mouth. Go ahead and chomp on that imaginary candy. Chomp, chomp, chomp, chomp, chomp. Swallow the imaginary candy and put your hand out, and I'll give you more imaginary candy.

Now, what happened? What did you notice? You salivated. As soon as your read "the tart, lemony interior of the candy," in an involuntary response, your mouth filled with saliva. When you chomped that candy up the intensity of that saliva increased, and as you swallowed it, you could taste the sugary exterior, you could taste the tart lemon interior. You noticed the physical response of salivation.

This is a great convincer for medical clients. Because what they do is say, "I'm in a really desperate situation, and I'm here because I heard this could be helpful, but I'm not sure how it's going to work." And when I do this convincer I can actually show them how it's going to work.

By the way, I never even call it 'real' candy. I called it 'imaginary' candy the whole time—but when we use the imaginative, the creative, the intuitive parts of the mind, and we put them towards a directed task, our bodies respond. In this case, your body responded by producing salivation from the indirect suggestion that you were tasting lemon candy. After I use this convincer, clients who have very serious medical conditions say, "Wow, that's incredible. I now experientially understand how this is going to help me."

The mind and body are not two separate things. When you change your mind or introduce new ideas to the mind it affects the body as well. These are the type of techniques that you'll learn in this course so that you can experience success with your clients.

If you have a burning need to understand the mechanism of hypnosis beyond what I explain here, read Ernest Rossi's book and the works of

Bruce Lipton. Also there's a one-hour lecture by Jon Kabat-Zinn, a researcher in the University of Massachusetts, titled *The Psychoneurobiology of Mindfulness Meditation.*

It's worth noting that meditation, self-hypnosis, yogic breathing—all of these things are really interchangeable in my opinion. These aren't ideas that are not researched. Hypnosis is not a complementary or alternative therapy. It is a well-researched and documented tool that can help you experience success with the clients that are on your caseload.

Hypnosis as a Teaching Process

I approach hypnosis as a skill-building process. Back in 2007, I did a presentation for the American Counseling Association's annual convention. It was an all-day seminar titled *Skill-Building Hypnosis.* Because I only had one day, six hours, I couldn't teach a complete course in hypnotherapy, so I taught them a basic strategy for induction, a basic strategy for deepening, and a re-orientation. I taught them no suggestive therapy. I didn't deal with direct or indirect or experiential suggestion; I didn't deal with convincers or anything else.

I taught a basic process for hypnotic induction, deepening, and re-orienting the client because *the process of hypnosis itself is a healing process.*

When I work with clients who come into my office, I view myself really as a teacher. I'm not doing anything to them; I'm not sprinkling magic pixie dust on them. What I'm really doing in the time that we have in my office is teaching them what they need to know so that when they leave my office or when hypnotherapy is completed, they'll be able to use these skills as resources for the rest of their lives.

My very first experience with a hypnotist was when I was eighteen. When I was eighteen I didn't know what hypnosis was, I didn't know was a therapist was, I didn't know what visualization was, I didn't know what NLP was, and I didn't know any blind people. But what I did know that was life was pretty stressful.

When I was eighteen, I'd had kind of a tough year. It was my first year in college, and everybody I knew had either died or moved. So I found myself my freshmen year in college with a weird living arrangement in this dorm with thirteen Puerto Rican baseball players. I kind of cut a deal with them: "I'll help you pass your classes if you let me camp out with you guys."

And I was also working; I was delivering pizzas for a pizza place. I was north of Chicago, and it was the winter time. I was driving my 1974 Pinto with a gray primer door, and orange and white stripes, and a lot of rust on it. Of course it was back before they incorporated safety procedures into pizza delivery. Some of you old timers remember: "Delivered in 30 minutes or it's free." So I was up against a wall with my pizzas in the back seat, and I was dealing with snow, ice, and rain; and of course I had a Pinto which was a serious handicap to anyone trying to earn a living as a pizza driver.

So things were pretty stressful, and a youth minister, who was kind of in charge of burger bashes and jello snarfing with the campus life group, noticed I was having a tough time. He said, "Look, you seem really stressed out, and we actually have a therapist on staff. I'd love it if you go see him." Well, because I respected and appreciated the fireworks and jello snarfing and burger bashing that the youth minister did, I thought "Ok, I'll go see the guy."

So I went to his office. Remember, at eighteen, I'd never met a therapist before, I'd never met a blind person before, I knew nothing about hypnosis, and I knew nothing about NLP, visualization, or any other technique. I remember sitting in the waiting room when he came out with this cane. He caned his way over to me and I thought, "Wow, I'm meeting my first blind person and my first therapist at the same time."

I didn't know what to expect, but he was very nice. He talked to me about my situation, the experiences I was having, and the stress that I was feeling. In retrospect, I recognize he without a doubt had training in Neuro-Linguistic Programming and certainly in hypnosis, based on the techniques he used in that session. And the session was really useful to me.

What he said was fascinating to me. He said, "Now do you do any visualization?" The blind guy is asking me if I do visualization. I found that fascinating.

I said, "I don't know."

He asked, "How about daydreaming?"

I answered him, "Yeah. I sit in class when I'm kind of tired, and I daydream all the time about the baseball players actually shutting up and letting me have a night of sleep. Or my pizzas getting there on time without the boss yelling at me."

And so he said, "I want to teach you a technique." That was his language choice: "I want to teach you a technique."

He said, "Kick back in the chair. And as you relax in the chair..."

He did a hypnotic induction. He didn't call it hypnosis, but he used a visualization technique where he had me use the imaginative and the creative capacity of the mind, and then he used an anchoring technique.

He said, "Now that you've created a state"--pure NLP--"of peace and calm and relaxation here in the office, I want you to imagine an old time camera in your right hand. The kind where you have a bulb you can squeeze."

He went on, "When you feel that state of serenity, calm, whatever you called it, just squeeze that camera as if you're taking a snapshot of it."

So I did.

He said, "Do you have a picture?"

"Yeah."

"Can you see it?"

"Yeah," I repeated.

"Bring yourself to that state again," he told me. "Take another picture."

Click. So I took another couple of pictures to look at.

And he gave me the suggestion that anytime over the next day or two or three while I was delivering pizzas, or in class, or with the baseball players, that I feel a sense of stress, anxiety, depression, despair--whatever it is that not the resource state—I should follow these directions: "Simply close your eyes for a moment, squeeze your hand together, and look at that picture you took here at my office."

I left his office almost thirty years ago, and yet I still remember that one-hour session with the therapist who without a doubt was a skilled NLP and hypnosis practitioner. I learned visualization, essentially, from a blind therapist.

What's amazing is I didn't just apply that the technique over the next three, four, or five days. To this day, thirty years after that one session, when I find myself driving—no longer in a Pinto—overwhelmed, stressed out, facing grief, or whatever is not a resource state to me, I still find myself squeezing that hand together and bringing myself back to that resource state with that same visual imagery that he gave me thirty years ago. In fact, the imagery that we created was so valuable to me that years later I actually bought all the furniture to re-create that imagery in my living room, so that my home became, when I walked in the door of my apartment, a center of serenity.

So the techniques that I'm going to share with you in this course are really global techniques. Strategies we can use with a wide a variety of clients, strategies that we should actually teach clients. Hypnosis is not something I do to people; it's something I teach them so that they can have this forever and ever and ever.

Chapter Two:
Basic Medical Hypnotherapy Techniques

Hypnosis is a pretty broad word. There are a lot of things that are hypnotic and are hypnosis. For example, a good car sales man is using hypnotic language patterns in conversational hypnosis. I'm a parent, and we know that parents, even if they don't have training in hypnosis, give us either positive or negative hypnotic suggestions. Hypnosis is a part of our everyday life and experience.

When I have clients who come into my office with medical conditions that need resolution, I view myself as teaching them the skills of hypnosis. And most of us associate hypnosis with "I'm in the comfortable furniture, I close my eyes, I count backward from ten, and now this is where the magic happens."

In truth, the magic actually happens as soon as they walk through the door, and it begins with me teaching them strategies and techniques. Some of these I use in formal trance. Some of them I incorporate into induction or into hypnotic suggestion. Some of these I simply share with my clients as techniques, so I might teach them in the intake room during the pre-talk and give them the assignment to practice a technique each and every day.

Autogenic Training

The first tool that I want to share with you is a technique called Autogenic Training. I'm going to explain two ways to teach it: during pre-talk, or in formal hypnosis as part of the induction process.

But first a little background on Autogenic Training. A few years ago, I wrote a book titled *Medical Meditation: How to Reduce Pain, Decrease Complications and Recover Faster from Surgery, Disease and Illness*. My friend Ziad Sawi, a board certified anesthesiologist, actually wrote the foreword for this book. I like to call it an airplane book. It's designed for the lay person to pick up, read through it in an hour or so, and learn some ideas that could help them. Autogenic Training is one of most important things in the book.

For some reason, Autogenic Training is not taught very often in the US. It is taught in the UK, it is taught in Australia, it is taught elsewhere, but for some reason in the USA, we just don't seem to talk about it very much. When I do see Autogenic Training talked about or taught in the USA, it is usually by nurses and in schools of nursing. You can read articles about the efficacy and value of Autogenic Training in some of the nursing journals. It was actually nurses that taught me Autogenic Training.

When I was finishing up graduate school, I was working at an in-patient psychiatric unit at a medical surgical hospital. Each night at nine o'clock, I would spend time doing relaxation and either hypnosis or meditation groups with all the psych patients. One day, one of the blue-haired nurses came over to me and said, "Richard, I have something for you."

She handed me a cassette, and it had a piece of masking tape on it that simply said "Autogenic Training" on it.

I said, "What's this?"

"Tonight when you do your group," she told me, "you'll probably really enjoy this."

Well, I had only a few different tapes that I would play for the patients in my psych groups. And now I had a new one.

So at nine o' clock, I used the Autogenic Training tape with the group that night. It was probably about twenty or thirty minutes long. I worked through the process as well, and when I was done with the process, I thought to myself, "That's got to be one of the most valuable techniques I've ever experienced."

I found that it centered me; it created a sense of wellness. It was just an incredible experience.

As I learned more about hypnosis, I came to realize that Autogenic Training is a process to actually elicit hypnotic phenomena—which is one of the most effective ways to help a person experience health, wellness, and healing. I use the hypnotic phenomena also in my basic hypnosis induction as a convincer. So there are many things that are really valuable about Autogenic Training.

What's even more valuable with it is that when we teach the skill of Autogenic Training to our clients and when they practice the skill of Autogenic Training, it lets our clients take control over their bodies. One of the biggest issues for the medical client who comes to see us for hypnotherapy is they feel a loss of control: "The cancer is controlling my body; the pain in controlling my body; my belly and my bowels and the accompanying sounds are controlling my body; the diabetes is controlling my body; my blood pressure is controlling my body."

My clients don't feel like they have control over their bodies. What Autogenic Training does specifically is teach my clients a process for reclaiming their bodies. And no matter what medical condition they come and present in my office for, this technique is truly useful.

I worked for William Rader's eating disorder treatment program many years ago, and I worked with anorexics and bulimics and bulrexics, and the accompanying medical conditions that they had. Autogenic Training was one of those strategies I taught, not only so they could feel physically better, but also because for the first time, they felt like they had control over their bodies in a way that was different than the maladaptive eating patterns that they have engaged in. So Autogenic Training is truly a useful and versatile strategy.

How did Autogenic Training come about? The history is really interesting. In 1932 there was a German cardiologist named Dr. Schultz. It must have been rough to be a cardiologist in 1932. This was before Plavix, before heart surgery, before all of the modern technology that we have in the field of cardiac medicine, so Dr. Schultz got tired of watching his patients die.

Finally, he said "Wow, all my patients die. There's got to be a way to help them not die." So he developed an eight-week training protocol he called Autogenic Training. Now *Autogenic* sounds like a fancy word, but it's really not. *Auto* means "self," and *genic* means "produced from within." So *autogenic* literally means "generating from within one's self." He developed an eight-week protocol that he taught to clients, so that at the end of eight weeks they would take physical control over the bodies. They would be able to regulate their blood pressure, their heart rate, and other aspects of their health and wellness.

By the way, the ideas of Dr. Schultz are consistent with the research Ernest Rossi, the ideas of Bruce Lipton, and the research of Jon Kabat-Zinn. He certainly was ahead of his time.

Now, you may be familiar with biofeedback, which became big in the 50's, 60's, and 70's. Everything that we do in biofeedback actually came originally from the ideas of Dr. Schultz. We can buy galvanic skin response meters or home EEG readers, or we can go to a biofeedback lab and attach all kinds of things to our bodies, or we can simply use the ideas of Autogenic Training: You have the ability to self-regulate your body. You can control your body, and you can learn to do that through practice.

Now we've all heard of yogis and swamis who can exhibit amazing control over their own bodies. Some can sit in a single position for hours; others can go without breathing for longer than anyone should; some can survive intense cold through visualization. Well, Schultz said, "Look, this is not just for the yogis in India. Everybody has within them the ability to control their own body."

Now sometimes I teach Dr. Schultz's complete eight-week protocol to my clients, but usually I'll teach an abbreviated version, basically week one of his protocol, so that's what I'm going to teach in this course.

Read over this process a few times to become familiar with it, and then practice it yourself. That's the best way to learn it:

Find a comfortable place. Take in a breath and close the eyes. As you breathe in and breathe out, scan your body. Anywhere you're carrying the tension of the day you can let loose that tension: the neck, the back, the brow. And of course continue to breathe in and breathe out, really focusing on this time of learning that you've set aside to increase your skills as a medical hypnotherapist. Let your hands rest on your lap and uncross your feet. Do what we would ask our clients to do in our office. Congratulate yourself for having made the decision to learn new things that you can truly benefit your clients. As you breathe in and breathe out, you can breathe in learning.

Now, focus on your hands. Notice how your hands are relaxed as they rest on your lap. And now with each breath your hands can become even more relaxed. Once you focus on your relaxed hands, notice how when hands are truly relaxed they're very heavy. Very heavy. In fact, as you notice your relaxed hands you can even say to yourself, out loud or in your own mind, "My hands are heavy. My hands are heavy." Go ahead say that, either out loud or in your own mind: "My hands are heavy. My hands are heavy." And let your hands feel that sense of heaviness as you breathe in and breathe out.

Think of the word warmth--warmth like that which might come from the sun or warmth like that which might come from inside of the body. Think of the word warmth, and as you focus on your heavy hands say to yourself, "My hands are warm. My hands are warm." And as you say that to yourself, notice the sensation of warmth in those hands. Actually say to yourself, "My hands are warm. My hands are warm." Notice the sensation of warmth.

In fact as you breathe in and breathe out, you can actually say to yourself, "My hands are warm and heavy. My hands are warm and heavy." Let your hands be both warm and heavy noticing that sensation. In fact your hands can be so warm and heavy that if you try to lift them you'll find it's impossible they're just stuck on your lap. If you try to lift your hands they're too heavy, simply stuck on your lap. Now, continue to breathe in and breathe out, with each breath doubling that sensation of relaxation, enjoying this time of learning.

Now reorient to the room. Simply open the eyes, feeling fantastic. Taking in a deep breath, filling the lungs with oxygen, letting that oxygen spread to every cell in the body, feeling fantastic.

That is a basic process of Autogenic Training. It's amazing how we can sense that feeling of heaviness, how we can sense that feeling of warmth. (If you didn't feel it the first time, that's okay. Keep practicing.) Some clients may feel warmth; other may feel heaviness; some may feel both. And of course, different clients have different responses. Some may feel very heavy, some may feel simply a sense of warmth, and others may actually feel hot. Nobody ever has the exact same response. I'll simply say to a client when I'm done with Autogenic Training, "Could you sense that feeling of warmth or that sense of heaviness?" They'll usually say, "I can sense both."

Now that's a basic strategy that's simply focused on the hands. But really that's an incredible process because your hands weigh the same now as they did during that process. And they weighed the same during the process as they did five minutes before I even introduced the idea. The sun did not beam into your room; nobody changed the thermostat on the wall. Your hands are probably the exact same temperature now as they were during the process and the same temperature as they were a few minutes before this process, yet you were able to generate from within a sense of heaviness, a sense of warmth.

Here's the question: If our bodies can create warmth, if our bodies can create heaviness, what other sensations can they create?

Can they create coolness? Burn victims could certainly benefit from that practice.

Can my solar plexus—that part of our body that is a nerve center for what in metaphysics we consider the "seat of the soul"–can it radiate healing to every part of my body? The answer to that is yes.

What other sensations could I create? If I can create warmth and heaviness, fairly simple things to do, why can't I decrease my blood pressure? Or increase the speed of my healing? Or accomplish any other goal by going through a process that teaches me to create from within that which is just going to be most beneficial to me and my medical conditions?

I teach a basic process for Autogenic Training to almost every client on my caseload. I might teach them only this skill. In some cases, I use it as a convincer instead of doing of the lemon drop because when my clients feel heaviness and warmth, they all say the same thing: "Wow, that was amazing."

When I'm doing a hypnotic induction, I usually incorporate some element of Autogenic Training. I may simply give them the suggestion that their hands are getting warm and heavy. I asked them to try to lift their hands, but gave the suggestion that they find the hands were simply too heavy. I've never yet had a client say, "Look hypno boy, it didn't work."

A lot of people think phenomena is for after the induction. Phenomena can be used during induction. It deepens trance. As a basic convincer, Autogenic Training both teaches a skill and has your client saying, "Wow, I'm really hypnotized. Even though I know I can lift my hands, they're just sitting there on my lap. Isn't that amazing?"

Now let's take a look at the extended version of Autogenic Training. This is basically week one of the eight-week protocol. There are several videos of me demonstrating this on YouTube, and of course, if you'd like the whole eight-week protocol, it's available on www.SubliminalScience.com.

As before, you might want to read through this a few times and then practice it yourself:

With your eyes closed, focus on your right arm and say to yourself (aloud or silently), "My right arm is heavy and warm." Just let that right arm be both heavy and warm. Say it to yourself, "My right arm is warm and heavy."

As you notice that sensation, focus on the left hand. Say to yourself, "My left hand is heavy and warm. My left hand is heavy and warm. My left hand is heavy and warm."

Now say to yourself 3 times: "My arms are heavy and warm. My arms are heavy and warm. My arms are heavy and warm." Notice the remarkable sensations within your body.

Next focus on your neck and shoulders, and say to yourself, "My neck and shoulders are heavy. My neck and shoulders are heavy. My neck and shoulders are heavy."

Then focus on your heart rate and say to yourself: "My heart rate is clam and regular. My heart rate is calm and regular." Say it to yourself, "My heart rate is calm and regular. My heart rate is calm and regular. My heart rate is calm and regular."

Now as you focus on your left leg, say to yourself: "My left leg is heavy and warm. My left leg is heavy and warm."

Focus on the right leg, saying to you, "My right leg is heavy and warm."

Say that 3 times: "My right leg is heavy and warm. My right leg is heavy and warm. My right leg is heavy and warm."

Now say, as you notice the experience: "Both of my legs are heavy and warm. Both of my legs are heavy and warm. Both of my legs are heavy and warm."

Now move your awareness to the solar plexus, what in athletics we might call the 'core.' It's that part below the sternum, above the stomach, the center of our being. In metaphysics this is of course a very important place in the body and say to you: "My solar plexus is warm and comfortable. My solar plexus is warm and comfortable. My solar plexus is warm and comfortable."

Now notice the forehead and the feeling of coolness on the forehead, saying to yourself, "My forehead is cool. My forehead is cool. My forehead is cool. My forehead is cool."

Repeat this short affirmation: "I am at peace. I am at peace. I am at peace. I am at peace."

Open your eyes, feeling wonderful

This is of course the basic protocol for Autogenic Training. I'll give this script, printed out on a piece of paper, to my clients, and I tell them to practice each and everyday between now and my next session with them. If your going to teach Autogenic Training to your clients, I'd recommend that you practice it yourself on a daily basis for at least a week.

Visualization and Primary Representational Systems

Visualization is another very important strategy for medical hypnotherapy. We use visualization often in hypnosis, and it seems like most of the hypnotists who I know consider themselves to be fairly visual.

According to Neurolinguistic Programming, different clients are going to be primarily visual, auditory, or kinesthetic—we all have different ways of learning. In fact, with most of the clients who come into my office, one of the first things I do is I give them a quiz that I developed titled the *Nongard Assessment of Primary Representational Systems.* (You can find a copy of it in Appendix A.) And when they're done with it, it helps me to determine what their primary representational system is.

(Autogenic Training is both kinesthetic, because it creates sensations and feelings, and auditory, because it uses self talk. So it's really a great technique for either of those.)

I sometimes describe it this way: Suppose you buy one of those assemble-it-yourself desks from IKEA. When you get home, do you look at the picture and then look at the pieces, and try to build the picture? That would be a visual person. Or do you just take a pile of pieces and start feeling, and looking and hold the screws, and compare the sizes? That would be a kinesthetic person. Or do you actually take the instructions and read through it step-by-step? That is the auditory person listening to the instructions. It's a pretty good test to find out your primary or dominant representational system.

I sometimes meet people who score a zero in visualization. It's not very often. About 60% of the people are primarily visual, but sometimes I meet people who'll say, "Yup. I just don't visualize."

This is one of those issues that hypnotists really struggle with. When we attach a specific meaning to a word and somebody can't reconcile the meaning we've attached with the word, they will detach themselves from it. Most people attach the meaning "picture something in your mind" to the process of visualization.

But visualization can really mean anything that we want it to. So if my clients say, "Nope, I just don't visualize anything. I close my eyes and don't see anything there." Then I ask, "What do you feel? How do you experience it? How is it interpreted?" So we can begin to use other angles to deal with visualization.

I also think, and this is an important concept for hypnotherapy, that when we have somebody who's not very visual, or not very auditory, or not very kinesthetic, instead of avoiding that with our client we should practice so they can develop visual acuity, so they can increase their auditory acuity, so they can become kinesthetically aware. The assessment of representational systems is not only useful so I can match their modality, but also so I can see where they need training.

So if I have a client who says, "Nope, I'm not visual. I just can't do it," I would probably start with a very basic exercise. I would say to them, "The research shows this is really useful, so I want to help you increase visual acuity." By the way, this is an indirect suggestion that as they increase their visual acuity through practice exercises, they will probably increase health as well so the research shows the correlation.

So I'll start with a very basic process. You can use any object for it, but I'll often use a singing bowl I have sitting on my desk.

Teaching a Client to Visualize

It's actually surprisingly easy to teach someone how to visualize. I usually do it in an interactive process, like this:

I'll say to my client, "Bring your attention to the singing bowl on my desk. See it on its colorful little mat there on the desk. You can see that, right?"

(Note: You can of course use any object. You don't have to use a singing bowl.]

"Yes, I can."

"And you see the different colors, correct? The green, blue, the yellow, the white?"

"Yeah, I can see that."

"And can you see the design? Can you see the pattern that's on the singing bowl?"

"Yes."

"Now go ahead and close your eyes. Do you know that the bowl is still there?"

"Yes."

(Note: I don't say, "With your eyes closed, can you see the bowl?" because that might bring resistance into play even though they probably can see the bowl since they've just closed the eyes.)

"How do you know that the bowl is there? I assure you I haven't moved it, so it really is there, but how do you know that?"

"I can sense it. It's as if I can still see it."

"If you were to reach out and touch the bowl, would you be able to do that with the eyes closed?"

"Probably."

"Go ahead and do it. Just touch the bowl even though the eyes are closed."

Once they touch it, I add, "What this means is that even though in the past you've had a difficult time with visualization, you could still judge distance and space, and have that extra sense. It's almost as if you have a sixth sense, to be able to intuitively know that the bowl is there. So a moment ago you were able to really involve yourself in the bowl, really look at it and see those colors. So ask yourself, even though your eyes are closed, can you experience that in any way right now?"

The client will usually say, "Yeah. It's almost as if I can see you."

Sometimes I'll use this line with client: "With your eyes closed, you're just looking at the bowl. It's almost as if you have X-ray vision and can just see through your eyelids. Can you see the bowl?"

The answer to that is yes, because I've given to them the hypnotic suggestion, just like the imaginary candy that produced a real response.

"Imagine you can see the bowl."

And so I can start very remedial with my client and go to exercises to increase their capacity for visualization. Hypnosis is a great tool for increasing visual acuity.

This is important because in addition to Autogenic Training, Visualization has a tremendous amount of research to demonstrate its efficacy in helping clients to achieve health.

Some Examples of Medical Visualization

Here are some cancer-healing visualizations that we can actually teach to clients:

Imagine an army of white blood cells, coming, attacking and overcoming the cancer. Imagine the white blood cells are like Pac-Man and they're going through the body and they're devouring up the cancer cells, or the unhealthy cells or the malignant cells, or whatever word we want to use.

Here's a great visualization idea:

Imagine a white light entering the body, bringing in energy, coursing through the body, then leaving the body and taking with it the bad stuff—the pain, the tension, the discomfort, even the cancer cells.

A couple of other great visualizations for cancer-healing:

Imagine white blood cells as knights saddled on horses riding through the body, attacking and destroying cancer cells. Imagine the cancer as small, easily squashed creatures being lanced by white knights on horses, white blood cells. Imagine goodness and purity in the immune system.

Here's one that simply deals with colors:

Cancer is a very dark color slowly turning paler until it's the same color as the surrounding body.

Another approach:

Imagine the cancer as being attacked by tiny bullets of energy. Weakened and dying cancer cells are being flushed out to the liver and kidney by your white blood cells.

Another visualization for cancer:

Visualize the tissue normalizing, being embraced by all the normal cells around it and counseled lovingly to change, to lose the prickles and the hostility.

Here's another visualization:

Imagine a sick little cancer cell, and imagine gently carrying it in your arms. Walking through a manor, where the wild flowers bloom, birds chirping, deer along the edge of the forest, coming to the edge of a cliff in the wilderness where you're able to drop those unhealthy cells you've been carrying.

Here's another great visualization; this is from Linda Lewandowski:

Cancer cells are misfits who do not belong, always fighting for more power because they're aberrations and they don't know how to love and they're lonely. So any decision I make about treatment I must temper with love. Love for them, no matter what they try to do to me. So envision little angels stroking them, blowing on them, sweet baby's breath to cool their fire, and take away their power.

Visualization is a strategy hypnotists usually are familiar with from the staircase deepener or from creating a scene of relaxation. Visualization can be combined with direct and metaphorical suggestion to help promote healing and wellness. Anytime I have client who comes into my office with any medical issues at all, I spend part of our time teaching them a skill like Autogenic Training, but I also spend time with them visualizing processes in the body succeeding and healing, recovering and feeling fantastic.

Future Pacing Health

Another Visualization process that I use often with my medical clients is Future Pacing. In order for anything to exist, it has to be an idea first. And so what I want my clients to feel when they step over the threshold of my door is that they're entering into their future. And that future is a vibrant and healthy, a bright place. So here is a script for a simple and short process that I use for Future Pacing success before medical treatments, procedures, and surgical processes:

Close your eyes. Breathe in, breathe out. And I'm going to guide you through a brief process using hypnosis and pain control. Bring yourself to the point that we call hypnosis. Breathe in, breathe out. Although your pain level may be high and you may constantly be aware of your discomfort, in time the treatments will be successful and come to an end. Go have that surgery, radiation, or procedure done. In the coming weeks, months, and years, you'll no longer be in the midst of the trials where you currently are. Instead you have helped solve your health problems. You'll look back on this time as a difficulty that passed.

So now, imagine yourself a year from now, seeing yourself as you know you will be. You're no longer healing but fully recovered. Allow yourself to experience the feeling of knowing you will have in a year. Experience the results that are important to you and know that your medical treatment is complete. Imagine yourself a year from now doing the things you currently can't do. See yourself bending, moving, getting up, getting out of bed with a wonderful feeling, maybe even running. Watch yourself doing things that may seem impossible. Involve your senses—sight, sound, touch—in this process. Picture yourself out of bed and outside smelling the scents of the day; and feel the pavement below your feet as you walk and hear the sounds of your neighborhood as you enjoy the radiance of the sun.

Of course we know that everything that is, was a thought first. And by creating this imagery, there's no question in my mind nor should there be in yours, that the future possibilities are endless. What the mind can conceive today, we know the body can achieve tomorrow.

Progressive Muscle Relaxation (PMR)

Over the years, far too many hypnotists have tried to impress other hypnotists with the "cool hypnotic secrets" they possess. They have, at some level or another, scoffed at rather simple approaches to hypnosis, especially progressive muscle relaxation, which many hypnosis instructors describe as out-dated, boring, or ineffective.

However, if you go to scholar.google.com—not www.google.com but scholar.google.com–and type in "progressive muscle relaxation," you will find literally thousands of journal articles demonstrating the efficacy of progressive muscle relaxation. This is a skill, which we can actually teach medical clients, that according to the research produces a *clinically significant improvement* in their functioning. Because of this, I'm a big fan of progressive muscle relaxation. It's an evidence-based approach to helping clients solve a wide variety of problems ranging from pre-surgical anxiety to pain relief. Much like Autogenic Training, PMR teaches clients to actually take physical control over their bodies so that they can in fact experience healing, health, and wellness.

Except in those schools that teach that PMR isn't cool enough, most hypnotists learn progressive muscle relaxation as their first hypnotic induction because it is a very natural and easy way to guide a person into hypnosis. So because of the value and the efficacy of progressive muscle relaxation as a skill, I most often use it as a strategy for hypnotic induction with my medical clients, regardless of what type of client. Whether a client who is seeing me related to childbirth, pain control, pre-surgical preparation, or one of many other reasons, I typically use a Progressive Muscle Relaxation induction.

Now, one of the other reasons why I use PMR is that I've never met my client before; they come into my office, and they expect me to help them. If I choose a different type of induction strategy, one that's much more "showy" or one that's much more "advanced," one that really shows off my skills as a gifted hypnotist, sometimes that client who's never met me before, never met a hypnotist before, can wonder: Is this what hypnosis is really like?

You see, a Progressive Muscle Relaxation induction is non-threatening. It's simple to use. I've never had anybody say, "Whoa, I'm relaxing. I'm totally creeped out. Stop that, Mr. Hypnotist." It's never happened to me. What they do instead is they say, "Wow. This is a valuable process. I feel both my body and my mind relaxing together." In short, Progressive Muscle Relaxation is a tremendously useful strategy with an impressive amount of efficacy and the research to back it up.

Now, most of us learn the passive form of Progressive Muscle Relaxation, which goes a lot like this:

And as you relax your eyes, you drop the chin toward the chest. And as you do that, notice the sense of relaxation in the neck and shoulders, and the relaxation spreading through the arms and hands and even into the little tiny muscles of the fingers. And as you breathe in and breathe out, allow that sense of relaxation to extend to the chest, and the muscles of the back, and buttocks, thighs, the calves, and even the toes.

That's a very quick example of a Passive Progressive Muscle Relaxation process, where we make the assumption that the client has the ability to respond to our direct suggestion that they relax those muscles, and that they'll do so. The alternate method is to actually teach your clients Active Progressive Muscle Relaxation, which teaches the difference between tension and relaxation.

In 1929 a physician named Edmund Jacobson wrote a book on Progressive Muscle Relaxation. Jacobson's approach deals with sixteen specific muscle groups. And Dr. Jacobson used this training protocol which taught the difference between tension and relaxation, where a client repeatedly tensed and relaxed each of those sixteen muscle groups during the course of PMR training. Now I have to admit although I have guided a class through all sixteen of Jacobson's muscle groups as an exercise in teaching them Progressive Muscle Relaxation as a hypnotic induction, I have never gone through all sixteen muscle groups with any client. That would probably be a thirty- or forty-minute hypnosis induction. It could have value with some clients, and I'm sure that other hypnotists certainly have used all sixteen of Jacobson's muscle group.

With most of my clients though, I use an abbreviated form of Active Progressive Muscle Relaxation during the induction process. And I do this so that my clients experientially learn the difference between tension and relaxation. Then, as I'm guiding them through the other muscle groups, they have the ability on their own to decide whether they want to actively tense and relax those muscles. And so, when I have a client who has an unusually high level of anxiety, or I have—we're going to put in layman's term, it's certainly not a clinical term—a person with a Type A personality, who I see as simply living in the high-stress lifestyle where they're not really comfortable with physical relaxation and probably don't spend much time doing meditation or relaxation exercises, I recognize that the idea of relaxation really is foreign to them. So I will often incorporate Active Progressive Muscle Relaxation into my hypnotic induction with those clients.

Here's how I do it. You will learn best if you follow this process while you read:

Kick back in your chair, set your hands on your lap, and take in a breath. And as you breathe in and breathe out, pay attention in this moment to the inhalation of the breath, to the way the air feels as you breathe that air into your nostrils and the way the air feels as you exhale from the lungs. And as you breathe in and breathe out, you've set aside this time to learn something new and valuable to you. So allow yourself with each breath to more fully enter that state of learning and that state of resource that we call hypnosis.

One way to experience hypnosis, of course, is to learn the keys to relaxing the body. As you relax the body, it's almost intuitive that the mind begins to follow. And as we relax the body, we'll find, after a few moments of time, both mind and body working together to achieve a state not of sleep, but of deep relaxation, where we can access that creative part of the mind. And so I want to teach you the difference between tension and relaxation. And as you breathe in and breathe out, scan your body and anywhere you notice that you're carrying the tension of the day, simply relax those muscles and let the body become relaxed.

And now focus on the hands; just pay attention to the left hand and the right hand. The right hand and the left hand. Although it may feel good to allow them to relax, I'm going to ask you to tense those hands by closing those hands into a fist, both of those hands. Close them up into fists as they rest on your lap and hold all the tension in those fists. Squeeze tightly, not to the point where you feel pain, but squeeze tightly so you can feel those fingers pressed into the palms and you can feel the tension of the skin across the knuckles and the fingers of the hands. Feel the muscles as you hold that tension in the back of the hands and even into the wrist and the forearm. And simply know what tension feels like.

Now release the fingers. Let those fingers open up. Let those fingers rest on your lap. Notice the tingly sensation of relaxation. Notice the difference between tension a moment ago and an incredibly profound sense of relaxation in those muscles at this moment. Notice the looseness of the skin, the heaviness of the hands, the feeling of relaxation.

As you breathe in, tense those hands into fists again. As you tense those hands into fists, feel the fingers in the palm of the hand, hold that tension, feel that tension, really grip those fists tightly, again not so tight that you feel pain, but feel it in the pinky, feel it in the thumb, feel it in the knuckles of the hand, the skin of the hand, the wrists, and even into the forearms. Notice the difference, again, between relaxation a moment ago and tension now.

And now relax again. Simply let those fingers open up; let those hands rest on your lap. Notice something interesting: The fingers feel twice as relaxed now as they did the first time we tensed and relaxed the hands. Some may even feel a sense of numbness, others a tingling sensation, some a sense of heaviness. All of us experience relaxation a little bit differently. But notice the profound sense of relaxation in the muscles of the hands, the wrist, the forearm as you relax the second time. This is, of course, the difference between tension and relaxation.

Most of us go through life spending our time unaware of our body's condition, neither noting if we're tensed or relaxed, simply going through the tasks of the day. But by practicing knowing the difference between tension and relaxation, you'll find that over the next couple of days when you feel tension in any part of the body, not just the hands, you can intuitively let those muscles relax and allow yourself to experience a profound state of mental and physical relaxation.

Now, focus on the muscles of the eyelids and the muscles in the brow, and the muscles in the cheeks. If you note any tension there, let those muscles relax. As you unclench the jaw, letting your jaw relax and the TMJ relax, with each breath double the sense of relaxation in both mind and body, feeling that relaxation spread to the neck and shoulders, and to each place in your entire body.

Now, if I were doing suggestive therapy with you, I would continue, giving your mind and body more and more suggestions for health and wellbeing, but since it's not, I'm just going to ask you to enjoy taking a breath, and as you take in a breath, note the oxygen feeling the lungs. Let that oxygen spread to every cell of the body, feeling alert and oriented and happy that you've learned through experience one more technique that could be truly valuable. Now, come to full alertness, ready to continue learning.

Now that was a simple process of Active Progressive Muscle Relaxation. While I generally use Passive Progressive Muscle Relaxation with most clients, with my anxious clients and certainly those "Type A" personalities who are always living in this high-stressed zone, I used an abbreviated form of Active Progressive Muscle Relaxation where they learn through experience the difference between tension and relaxation by using that process in the hands and in the fists.

Again, Progressive Muscle Relaxation is an evidence-based, an outcome-based therapy. It is something that as a hypnotists we should not discount because of its simplicity. It is actually something we should use. It teaches a specific skill; it comfortably guides a person into a deep trance state. And it's a process that many of my clients, even if they aren't familiar with hypnosis, are actually familiar with or at least have heard is beneficial to them. And so it causes them to respond, in the same way developing rapport does, with a sense of openness and willingness to engage in the other hypnotic processes that we utilize.

A Word about Deepening Techniques

After we've done a hypnotic induction, we generally do a deepening procedure. Although this isn't a course on deepening—and I'm making the assumption that you have a basic knowledge of the hypnotic process—I'd like to discuss my favorite deepener.

Again, I like to keep things simple, and so my favorite deepener, although it's certainly not the only one I use, is a simple number count. It's consistent with my client's expectations. They've seen hypnotists count backwards in Hollywood movies. And it's something that I can come back to as a transitional deepener throughout the session.

Here's how it goes:

As you breathe in and as you breathe out, with each number and each breath allow yourself to double that sensation of relaxation. Five, four, three — you're doing perfect—two, one, zero. All the way down now. Completely relaxed.

I recently worked with a client who was a corporate executive, an athlete who really lived in a high-performance zone. I recognized immediately that managing stress and creating a sense of serenity and calm was probably not his area of expertise. So I used an induction that included Active PMR, and I'm very happy that I used the number count deepener.

You see, throughout the session he would periodically adjust for comfort and move to a higher level of alertness during the process. This is to be expected. Rarely do I have a client who goes into trance and spends eight hours there. All night long we tend to go up and down and not worry about trance depth too much. But if I noticed that at any point he was getting a little bit too oriented, or I noticed at any point he seemed to become distracted by following his thoughts, I was simply able to say to him, as a transitional deepener, "and as you continue to relax five, four, three, two, one, zero," and then I would give him more suggestions. I would give him a story, a parable, a metaphor, or an indirect suggestion. I would see him begin again to almost cycle like a person does during the evening of sleep, becoming more oriented, and I would simply say, "and breathe in again five, four, three, two, one." And so we can use that number count as a simple transitional deepener coming back to it throughout an entire hypnosis session.

Time Distortion

Almost every hypnotist is familiar with Time Distortion. In fact in 1954, Milton Erickson wrote a book with another physician named Linn Cooper, and that book was titled *Time Distortion in Hypnosis.* It is one of the observable and frequently occurring phenomena of hypnosis. I almost always end a session by using Time Distortion as essentially a post-hypnotic convincer.

Here's how I do that: When we're done with the session, I say, "And so, about how long did that session feel for you?"

They'll almost always say, "I don't know." And probably eighty-five percent of them say it felt like two minutes or five minutes, or sometimes ten minutes. Then I'll point to clock and say, "Actually I was timing the session; it was twenty-eight minutes," or however long the session was.

They'll usually respond, "Wow." And so my response to them then is, "What that means is that you were really able to let go, involve yourself in the process. We call that Time Distortion. It felt like five minutes but it was almost thirty minutes, and so that's a great thing. That lets me know you're going to do really well when you head home today."

About ten percent of my clients say, "Wow, well I know I couldn't have been here two days, but that felt like two days." Or they'll say, "I know couldn't have been here eight hours, but boy, that felt like eight hours."

That's Time Distortion in the other direction, making thirty minutes feels like thirty hours. And I'll say to them, "Well, of course, you weren't here for thirty hours. You were only here for twenty-eight minutes. What that means is that you were really able to let go of time and really involve yourself in the process. And so when you head home today I know you're going to be one of those who does extremely well in response to our session."

So Time Distortion, either way, can be a great convincer at the end of the session.

You may be saying to yourself: What about the other five percent? Here's how to handle it: I'll say to them, "Wow, that's awesome. A lot of clients think it feels like five minutes, or some them think it feels like eight hours. But you were really able to pay attention to the entire process. It was about thirty-two minutes, and you said that it was about thirty minutes, so you're right on the money. That lets me know that when you leave our session today you're going to do fantastic because you were really able to follow the process and internalize the suggestions that I gave." They'll have a big smile on their face just like the other ninety-five percent, and they'll say, "Thank you very much."

It really is a great convincer because when I make them aware of the presence of Time Distortion, then they know that something happened. And of course if something happened, then something happened—so it becomes a proof to them of the validity and value of the process.

Note that all this happens without you having to suggest it. That often has the client wondering if you did suggest it without them remembering it—which serves to compound the effectiveness of your convincer.

Most hypnotists are not using Time Distortion as a phenomena in hypnosis any other way. So I'm going to share with you a technique that has a specific application to medical hypnotherapy for the clients that we work with. A lot of clients that we work with either have chronic or acute conditions, so they're going to spend a lot of time waiting in the doctor's office.

A patient might show up at 1:45 for a two o'clock doctor's appointment and end up waiting until 3:15 or 3:30. Agitation, frustration, and more importantly rumination and fear can really kick in during those waiting periods involved in medical treatment.

There's another area of medical treatment where our clients really suffer because of time, and that is the duration of certain tests and procedures, such as MRIs and CT scans. For example, it's not uncommon for a client to be put inside of the machine and to spend a period of time in there. It may only be three to five minutes or it may be a much longer period of time, but it becomes a very anxious time for them. It becomes something that they have a hard time enduring. They experience negative time distortion.

And there are other procedures of course where our clients are told, *"You must swallow this, or you must do this, now you must wait for this."* During these often very uncomfortable procedures and processes, our clients can manifest a tremendous amount of anxiety, rumination, fear, catastrophization. So if we can use Time Distortion as a tool for post-hypnotic proof, we can also use Time Distortion as a concept for helping our clients during difficult time periods: during the surgical process, during the diagnostic process, during the waiting process, to experience comfort, relief, and wellness.

Time Distortion is also a great tool for dealing with phobias. A lot of our clients really are fearful and phobic of medical processes, medical procedures, diagnostics tests, and other things.

Let me give you an example from my own life. I used to, a long time ago, be afraid to fly. I'm now an Executive Platinum Flyer on American Airlines. I fly very often. In fact I don't even like driving anymore. If I could fly between my office and my house, I would, because flying is just more comfortable for me. It's the way I like to travel. But there was a time when I was afraid to get on a plane and one of the things that helped me overcome that phobia was the use of Time Distortion in self-hypnosis. I would take an eleven-hour flight from Dallas to Tokyo and make it feel like thirty minutes or an hour—or an eight-hour flight from Dallas to Frankfurt and make it feel like an hour. And so when I started doing a lot of international traveling about a decade ago, I really used the idea and the concepts of Time Distortion to help me manage fear and anxieties as well as to simply help me be more comfortable during the process.

Our medical clients may not be flying, but they may find that they have a fear of the CT scan, for example–a lot of people do. That fear can be alleviated by using the concept of Time Distortion.

Here is a simple script for Time Distortion in hypnosis. Use whatever induction works with your client and then say:

Now imagine a clock or think of your watch. Imagine it; both of the clock's hands are at twelve o' clock. Imagine seeing that clock and realizing that it's no longer important what the numbers on the clock say. In fact, the hands of the clock could be pointing at six o' clock, or three o' clock, or twelve o'clock; it wouldn't make any difference to you. And as you continue to relax you can let go of the meaning of time and let your clock move from twelve o'clock all the way to seven o'clock. And imagine it moving from seven o'clock to one o'clock. From one o'clock to five o'clock. And from five o'clock to nine o'clock. You can imagine the hands moving together as one. Or you can imagine one hand moving faster and one hand moving slower.

You can also imagine that the hands of the clock not only go clockwise but can even go counter-clockwise. No longer concerned about day or night, afternoon or evening, three or six, seven or two, or eleven and four. And you can bring yourself to this time and this place right here and right now where time is unimportant. Notice the feeling of timelessness. Notice the process of inducing Time Distortion. Notice the value of simply letting go and using that creative part of the mind. It really feels wonderful to access this state, this resource state that we can utilize or that we can teach our clients. By going through this process undoubtedly you can, not only with the intellectual part of the mind, but now also the experiential part of the mind, understand the value of this process.

This is, again, a fairly simple process for Time Distortion, but a powerful process for Time Distortion, because as people in the modern era we are programmed to attach meaning to time and to see times sequentially throughout the day. When we use the imaginative part of the mind to see time no longer going in the course or direction that's consistent with our expectation, we can let go of time.

By the way, this is not only a great way to induce Time Distortion, but it's also a great strategy for letting people who have difficulty setting aside time for themselves to begin that process of self-hypnosis, so that they can actually take time for themselves during what many clients report are busy days.

When I have a client who says, "You know, I'm really too busy to practice hypnosis." I often respond, "Well, then you don't really don't want to experience change, do you?" Confrontation can actually be a pretty valuable.

But I'll sometimes have client who legitimately has a packed day, like a single mom, who's taking care of little kids, all those sorts of things. The days are tight. And they'll say, "I don't have time for twenty or thirty minutes of practice for self-hypnosis."

So I'll say to them, "Unwind from the importance of time, setting aside for yourself a space for taking care of yourself, whether it's five minutes, or ten minutes, or twenty, or thirty, letting go absolutely by beginning your self-hypnosis time with this pattern or this script in your self-talk."

This is a truly useful process. Practice it yourself so that you can use it with the medical clients that you work with.

Radical Acceptance

Radical Acceptance is an essential strategy for medical hypnosis. Although it would be nice if all of our clients who have medical problems achieve perfect health and immortality, essentially discovering the hypnotic fountain of youth, never again to suffer any ills, any problems or any difficulties, we all know that people do not live forever because there comes a point where our bodies, our earthly vessels, will succumb to the inability of life to be sustained forever—or at least in mortal terms.

And so, we have to deal with the issues of acceptance. Many clients come to me with unchangeable medical difficulties. There is nothing I can do and nothing the doctors can do. And I am a person who does believe in miracles, and I have certainly seen some miraculous things, so I would never begrudge a client for holding out hope. But I think it's important for us to recognize that sometimes medical situations are in fact tragic. And one of the most effective interventions is not providing false hope, but actually encouraging acceptance.

Now acceptance does not mean you like something. Acceptance does not mean you endure something. Acceptance doesn't mean you hope this happens to others. Acceptance simply means that you are at a point where you can lay something out, look at it, and see that it exists, even if you have disdain for it—and then move on from it. That is what acceptance is all about.

The paradox of acceptance is that acceptance gives us a sense of freedom. In life we cannot avoid pain, but we can avoid suffering. Suffering often comes about because we fail to accept. And so a big part of medical hypnotherapy is working with clients to accept those things that they want to find unacceptable.

I shared my experience with my foot surgery. When I had the foot surgery, I perhaps believed that it would solve all of my problems. While it resolved the problem of the arthritis pain and the degenerated joint, the solution brought about a new set of problems. I guess I really wasn't prepared for those new problems to be there. But guess what, those new problems are never going away. Why? Because my foot has metal rods and screws in it. There's no way to put the joints back in. There's no way to reverse the surgery. What is, is. I'm actually happier with the outcome the way it is, but I find certain elements of my foot unacceptable. After my foot surgery, I realized that I would now have certain limitations, and I really was angered by that. You see, anger is a relational frame. It's something that I've attached to that pain that has created suffering. And the process of acceptance brings freedom.

The paradox is that when something is accepted as just being what it is, then it holds no power. Depression, loneliness, hunger, pain, hurt, injury, illness, or even withdrawal become unimportant when accepted. When something becomes unimportant, it becomes just 'what is'; it is something experienced rather than something I fight, or something I hate, or something I'm restricted by or obsessed with. And I can find freedom from suffering. Happiness would suck if life had no depression, because we wouldn't know what happiness was. Security would suck if we had no sense of fear to put it into perspective. Difficult times and experiences are a part of any valued path. Difficulty and pain are not to be avoided if one wants a truly meaningful life. In acceptance we give those things no power to control, and those things become the pathway of being a full human being, participating fully in life.

When working with clients with medical conditions, I will often have them go through a process of acceptance. Here is that simple but profound process I often go through with clients:

Imagine sitting in a conference room table, just you. (Of course you can create this visualization anyway you want it to; you can simply see yourself sitting in a field alone.) As you breathe in and breathe out sitting where you're sitting, imagine lying out in front of you either on that table or on the ground is that which has been so distressing to you or that which you've been unable to change or that which has caused you pain. Simply see it, look at it, and observe it. Note every feeling about it you have. You don't have to stop being angry or stop being hurt. But don't allow yourself to become immersed in those feelings either. Simple noting them, letting them come, letting them go. And looking, observing, seeing that which is so distressing to you simply in front of you, maybe it's your cancer, or your injury or your pain. And take as much time as you need to simply observe it. To note that it's there. You don't have to like it, endorse it or want it to happen again, but simply observe it and allow yourself to be with it, noting that in acceptance of its presence we find freedom from our suffering.

I would of course give the client as long as they needed, possibly exploring this in more detail, but this gives you the essence of the exercise.

Emotional Freedom Technique (EFT) and Acupressure

While both of these techniques can fill volumes, I'd like to briefly talk about their value in medical hypnosis.

Both approaches are based on Chinese medicine and operate on the same principles as acupuncture. There is considerable debate as to whether these work because of unseen energy centers in the body, some function of the nervous system, or simply the power of suggestion and ritual. It's not the purpose of this course to sort that out, but merely to point out that there are many studies showing the clinical efficacy of EFT to reduce distressing symptoms.

So I teach EFT to some of my clients. This course is not really designed to teach you EFT, so I encourage you to find a suitable training program to increase you skills in this area of intervention.

Chapter Three:
Suggestive Therapy for Depression

Understanding the Depressing Triad

During times of medical difficulty, depression is certainly something that we're going to deal with as hypnotherapists. Many people come into our offices because they would like to learn the strategies and skills that can help them recover, experience healing, and address specific issues. But almost any kind of client who comes to me for any type of medical condition also wants to feel a sense of relief from the depression that surrounds the medical problem that brought them to my office in the first place. So a very basic skill we can develop as hypnotherapists is how to counter what I call the Depressing Triad.

The Depressing Triad is the set of actions that we take in response to our beliefs and the feelings that are generated by those actions. Many of my clients actually have a fairly unhealthy or limited set of coping strategies. So if they believe unhealthy things about themselves, their future, or their medical situation, they almost always take action on those beliefs. The actions that they take are unhealthy choices: overeating, drinking, drug abuse, over-use of prescription drugs, isolation, withdrawing from people who care about them, not sleeping at night—really denying themselves any effective coping strategies.

When your clients believe negative things to be true about their situation, they will act in those beliefs. And those unhealthy actions can in fact exacerbate the situation, so it becomes a situation where depression leads to despair, despair leads to hopelessness, and hopelessness—at its extreme—leads to suicidal ideation or self destruction through maladaptive coping strategies, which can lead to death. And of course by not being focused on health, we're focused on sickness, which can dampen the body's ability to manage immune system responses that help a client to experience healing. So dealing with this depression is really important.

The first element of the Depressing Triad is the situation. The second element of it is how we see our self. And the third element is how we see the future. The Depressing Triad is a set of what some would call "cognitive errors" that our medical clients develop during the process of illness. Because these cognitive errors can really exacerbate the client's inability to experience recovery, I'm going to share with you a specific hypnotic technique that actually counters all three elements of the Depressing Triad.

The first element of this triad, the first area of cognitive errors, is the situation: A person feels ill, they notice a problem, a bump, a lump, or whatever. They go see the physician, and the doctor says, "Hmm, this might not be good. We're going to need further tests."

Our client's first thought is, "The doctor did not give me an answer; that means that this situation is awful."

And so they catastrophize a situation, viewing it as tragic, viewing it as awful, viewing it as unrecoverable, viewing it as terminal. And so the first element of belief that we're going to be dealing with are the catastrophizations surrounding a specific situation or the devaluation of that situation. That's the first element of the Depressing Triad.

Now all of these are normal human responses: normal people experience this, not weak people or inept people or people who don't love God or have some moral or intellectual failing. All people experience to some extent these normal responses. But just because they're normal responses, that does not necessarily make them healthy responses. As hypnotherapists, we're going to recognize these normal but unhealthy responses, and we're going to help our clients develop health and wellness even in the midst of these catastrophic situations.

So the first thing I do when I'm talking to a new client is spend time really looking at how they characterize and perceive their situation. What absolutes or what meanings have they attached to this event? How have they catastrophized it?

That's a vital first step, because left unaddressed, the client's thoughts progress to a point where they not only devalue the situation, but move into a devaluation of the self, thinking:

"This situation is awful; I must have done something wrong. This situation is wrong; I shouldn't have done those things that were enjoyable but bad for me. This situation is awful; I'm not strong enough to handle it. This situation is devastating; I'm alone and isolated, and nobody cares about me. This situation is intolerable; therefore I have no value."

So you can see the progress from devaluing the situation to devaluing the self. Again, these are cognitive errors, beliefs that are not true.

Now after devaluing the situation and devaluing the self, if we haven't yet created an intervention that assists the client in changing their response, the natural progression is the third point in the triad: They devalue the future, thinking,

"Therefore, I'm going to die. It is hopeless."

And so this Depressing Triad comes about with a lot of our medical clients, particularly as we deal with second reoccurrences and particularly when we deal with corrective surgery in relation to the first surgical process.

And so, I have a client who might say,

"Wow. The cancer is back." That's the situation. "This is intolerable. This means I must be one of those people who isn't going to recover."

Then they devalue themselves and say,

"I must have done something wrong. I'm not ok; I should have taken better care of myself," or they voice whatever self doubts that they have.

And then, if left unaddressed, they move into that future orientation where they see themselves as dead, sick, or physically limited in some capacity, rather than seeing themselves as well.

And of course, we know that whatever the mind attends to, it considers; and what it considers, it eventually acts upon. And from a mind that attends to a future that is hopeless, my clients will act hopeless. So for all of the medical clients whom I work with, one of my first interventions is to reverse the process of the Depressing Triad.

Countering Depressive Self-Talk

A great strategy for halting the Triad is using affirmations and self-talk. Therapists traditionally call this "countering cognitive errors." It can be done in a number of different ways. For example, a non-hypnotic way that I do this is to give almost every client I work a dry-erase marker. If I have a client who is catastrophizing, who's engaging in negative self-talk, who's telling themselves things that aren't true, I give them that dry-erase marker with the assignment to take the dry-erase marker home and to write on the bathroom mirror the counter of their cognitive error, or write down the opposite of whatever it is that they believe in this Depressing Triad.

For example to counter the idea that their situation is hopeless, they might write,

"The truth may be this situation is difficult, but people do recover. I am a people, and I can recover."

For catastrophizations of the self, the cognitive error might be,

"I'm not a worthwhile person, and there must be something wrong with me."

The counter to that cognitive error may be,

"I am a creation, a spark of the divine, and within me is the power to heal."

Maybe they move the point where they devalue the future, and they've been saying to themselves, "My future is hopeless."

I might have them write with their dry-erase marker on the bathroom mirror the counter to that cognitive error, which is:

"As long I am breathing, I am okay. I will continue to breathe each and everyday."

Whatever the cognitive error, there is a counter, a positive affirmation.

46

The reason why I have people write on their bathroom mirror is really pretty simple: It's the first thing they see in the morning. It's usually the last thing they see at the end of the day. They can actually still blow dry their hair and do whatever it is that people do in the bathroom, and it becomes a self confrontation to them, these counters to their cognitive errors. So even if they don't believe it's true when they write it on the bathroom mirror, at some point they're going to see those words over the next couple of days and weeks, and say, "I wonder if that's true." And when they wonder if it's true, they will take action to test it. That is human nature. When they take action to test it, they will find that what they believed previously was untrue, and they'll replace those cognitive errors with the truth.

Hypnotic Processes to Counter the Depressing Triad

Now as far as hypnotic methods go, hypnosis is a great tool for essentially re-writing the scripts that we have created in our subconscious mind. It is a great way to access the Tabula Rasa, the clean slate, and to engrave a new message to our self. And so I use a hypnotic process for this.

Here is a script of that process as I use it for my weight loss clients:

As you relax in the safe place of this office, pay attention to the feeling of depression you have often felt. But this time, pay attention to it in a mindful way, seeing that depression you feel just as depression, intentionally choosing not to attach meaning to it. Simply acknowledging it. It is just depression, a feeling you are familiar with, but in the safety of this office you are just noting it and looking at it. Anytime you catch yourself saying "this is why I am depressed" or "that means I should do this or that" just return your focus to that of an observer, just looking at or feeling or even listening to the voice of that depression without judgment.

Often depression develops because we find ourselves in a situation that we have labeled as depressing. For example, you recently (insert behavior here, these are just examples) *promised yourself that you would not eat sweets but found yourself eating dessert every night for several days. It may be habit to say to yourself, in a situation like this, "I can't help myself; I will always fail at my goal." And surely you have been telling yourself statements like this which devalue your situation. Have you ever noticed that left unchecked these thoughts naturally give birth to a new thought, one that devalues the self? With statements like "I will always be fat" or "I am ugly" being the natural order of this unhealthy thinking? Again, you have certainly found this pattern being one that eventually has led you to believe and say to yourself, "I will always be alone" or "I am unlovable." In this time of hypnosis, you can become an outside observer, even imagining floating outside of yourself to a new vantage point, where the real you is able to look at the first you, and seeing this process being a pattern of feeling and thinking. Do you see how thoughts give way to more thoughts, and the process, if left unchallenged, brings more thoughts? It is almost like a waterfall, where more water simply follows the previous water and seems to have more power and pain with each rush of the stream. As an outside observer though, rather than being one who is under this waterfall, you can see the source of the water, the steam or river at the top of the waterfall feeding these thoughts one upon another.*

The way to turn off the water is to change your self-talk, challenging these automatic thoughts. You have already learned the process of mindfulness, and so by practicing mindfulness you can take a break from feeling like you are in the waterfall and become an observer instead. By floating outside of yourself and seeing the first you under that waterfall you gain a new vantage point and can counter these cognitive errors with the truth. And of course, the truth may be painful, but pain is far different than suffering. Pain is merely a signal or an observation rather than suffering, which is the judgment we create of our pain. And of course, it is also entirely possible that from this new vantage point, you will also notice the presence of more than pain, the feeling of comfort, safety or even security. And so look at the facts of the current situation, telling yourself the truth:

Situation: "I find sweets tempting and sometime make unhealthy choices, but I still have my choices"

Self: "I am not my weight; I am more than a number"

Future: "Even though I may not make progress as quickly as I expect, I learn from my experiences"

Here is how I adapt that process for medical clients:

Go ahead and close your eyes. That's probably the most effective way to learn—and if your eyes closed and you find that really enjoy hypnosis and you don't pay attention to each and every word, that's okay. Or you can pay attention to each and every word I use. Really, either way is fine with me.

As you relax, you can pay attention to that feeling of depression you are facing along with the medical difficulty. Pay attention to your feeling of depression in a mindful way, seeing it is just as a feeling, intentionally choosing not to attach any meaning to it. Simply acknowledge that part within you that can feel depression. In the safety of hypnosis, you can view it, simply note it, and if you catch yourself saying, "This is why I'm depressed" or "that is why I'm depressed," simply draw attention to your breath, not attaching any meaning to that depression, simply letting that depression exist, almost as if you are an observer of that depression. Often depression develops because we find ourselves in a situation that we've labeled as depressing.

For example, you recently discovered a medical difficulty and you promised yourself that you would do what was necessary to take care of yourself, but in this situation you may have found yourself saying, "I just don't have the energy to succeed. I can't help myself. I'm going to fail." And surely as you've been telling yourself statements like this, which devalue your situation, left unchecked, you will begin to devalue the self with statements like: "I am not ok. I am not well." Or even statements like, "I deserve this illness." Perhaps you've noticed that eventually these thoughts move on in their progression, so that as you look at the future, you might have said, "I am unlovable" or "I am alone" or "the outcome is tragic." But fortunately during this time of hypnosis, you can become an outside observer. You can even imagine floating outside of yourself to a new vantage point, where the Real You is able to look at the First You and see this process as being a pattern of feeling and thinking.

As an outside observer, do you see how thoughts give way to more thoughts and the process, left unchallenged, brings more thoughts? It's almost like a waterfall where more water simply follows the previous water and seems to have more power and pain with each rush towards the stream. But as an outside observer though, rather than being one who is under the waterfall, you can see the source of the water, the stream or the river on top of the waterfall, feeding these thoughts one upon another. The way to turn off the water is to challenge these automatic thoughts to turn off the self-talk. Because you've learned the process of mindfulness, by practicing mindfulness, you can take a break from feeling like you're in the waterfall and become an observer instead. By floating outside of yourself and seeing the First You under the waterfall, you gain a new vantage point and can counter those misbeliefs with the truth. Of course the truth may be painful, but pain is far different from suffering. Pain is merely a signal or observation rather than suffering, which is a judgment we create of our pain. Of course it's entirely possible that from this new vantage point, you'll also notice the presence of more than pain, the feeling of comfort, safety, or even security.

So as you look at the facts of this moment, tell yourself the truth. The truth is, of course, that the body is a wonderful machine with a built-in program for healing power. The body is a wonderful machine with a built-in program for healing power. Say that to yourself as you continue to relax. Say it in your mind or say it out loud, "The body is a remarkable machine with a built-in program for healing. The body is a beautiful machine with a built-in program for healing." And as you say these things to yourself, step into the truth of those counters to cognitive errors.

And so, take in a deep breath. Being ready in a moment to open the eyes, letting oxygen fill the lungs reaching every cell of the body, when I count to three, opening the eyes feeling awake, alert and oriented. One, two, three. Opening the eyes, and of course if the eyes are not open yet, take in a deep breath and open the eyes.

I wanted you to see both scripts so that you'll be able to see the adaptations that I made. You can adapt the script to the unique needs of your specific clients. This is a great process, a great metaphor; it uses the strategy of dissociation, it uses the metaphor of the waterfall, it counters the cognitive errors, and it uses affirmations. It's an effective hypnotic process that's truly useful for dealing with the depression that we see among most of the hypnotherapy clients that we work with who present with medical difficulties.

Chapter Four:
Pain Control

Pain Control Principles

Probably one of the most frequent reasons for clients to seek out a hypnotherapist related to medical conditions is to deal with pain—usually chronic pain, but sometimes acute.

Acute pain is a pain that is not usually present, but has emerged either because of an unknown or known event that is causing a client distress. For example, if I were to punch a door with my hand, that would cause acute pain. My fingers will hurt.

Chronic pain will happen if I get my hands stuck in the door and break those fingers—if they don't heal quite right and I develop arthritis in those joints, I could feel an ongoing pain in those joints for years. That's chronic pain.

Now, pain always serves a purpose. The purpose of pain is to tell us something isn't right. We need to do something to become physically congruent again, to be restored to a state of equilibrium. In the case of pain from a broken bone, the purpose of the pain is to say, "Hey, don't put this off anymore. Go to the doctor, get the bone set, and get a cast."

The purpose of chronic pain really is, in many cases, to tell us to take care of ourselves, to find new ways in adapting, to take action. It can help us to acknowledge, not only from a physical perspective, but even from a spiritual, emotional, or social perspective, to make changes that are going to allow us to live fully in this moment and in this world.

The processes that we use with pain control can be used with both acute and chronic pain.

If a client presents with issues related to pain, there are two things that are paramount here:

1. Pain is the reality of the client's subjective experiences.

I love the nursing definition of pain: "Whatever the clients says pain is, is what pain is."

Our clients have already had a professional doubt whether or not what they're experiencing is valid or true. People see a hypnotherapist last, not first, so they've already been discounted by medical professionals as well as other professionals, and the last thing they want to here is somebody who invalidates their experience.

Now it may be true that the client sitting in front of you has a much different interpretation of pain or pain threshold than you do. And if you were experiencing what they are experiencing, it may be true that you would have simply let it roll off your back, that you'd continue to go. People seem to have different pain thresholds. It wasn't too long ago during the Summer Olympics in London that one of the runners in US relay team actually broke his leg during the relay race, and yet he continued to race. Well that's a much higher threshold of pain that I'm ever going to have. So If I said, "Ow. My pinky hurts," he might say, "That's not real pain. You haven't felt real pain before. Real pain is when you run a relay with a broken leg." I am not the judge of the quality of my client's pain. I'm a hypnotist; it's my job not to discount their experience. It's my job to provide them solutions to whatever they are calling pain.

2. Pain is the signal of a medical condition.

There are certain types of pain that need an immediate medical response. Because I work with men and some of those men have sexual issues, one of the issues I encounter, of course, is pain in the testicles. Any time a client reports that, it is time for an immediate evaluation by a urologist. There are many other types of pain that clients may manifest, and I need to recognize that as a hypnotist, I may be skilled at reducing the severity of a client's pain or subjective experience of pain and helping them to function, but sometimes, pain serves a purpose.

Let me give you an example from my own experiences of how hypnosis may not be so helpful at times. Years ago I developed arthritis in my foot. And that arthritis began to do what arthritis does and disintegrate the joints. Now, because I'm a hypnotherapist, because I'm a busy guy, because I don't want to take time out for surgical processes and those sort of things, I used my background and personal training to assist me in developing physical strength. I used my skills as a hypnotist to modulate my pain, even in scenarios and situations where people may have been using pain medication or even have become immobile. And so for years, I put off addressing, medically, the pain because of my own ability to use effective interventions to delay treatment.

Unfortunately by the time I actually scheduled the surgery I'd been told five or six years before to have, the joint had completely disintegrated. Instead of being able to do a joint replacement, they ended up doing a fusion. Now a replacement would be been preferable, but because I did not earlier address what the pain was telling me, we had to do a more severe surgical process with a less desirable outcome. Instead of taking care of myself, I simply masked the symptoms using the techniques that were helpful to me. Now fortunately I also used those same techniques to experience rapid healing and to exceed my doctor's expectations about the outcome of the surgery.

For example, the doctor told me I would never run again. When I go to the gym, one of my favorite things to do is, of course, run on the treadmill. And he told me I would never be able to do that again. But using the principles of techniques of healing, mindfulness, and self-hypnosis, I have certainly exceeded the doctor's expectations for what a person with the type of surgery I had would ever experience.

But my point in telling you this is really pretty simple: Let the medical professionals make those decisions. The orthopedic physicians, the oncologists, the other doctors who are treating our patients need to make the decision about the type of intervention that is necessary.

When I teach these techniques to my clients, I want to always provide my client with the admonition that this is not a replacement for medical treatment; this is tool and a resource to help them experience a sense of fulfillment, a decrease of symptoms that are distressing to them. They should continue with the traditional medical treatments that their physicians recommend. When we are treating pain clients, it's essential that your client has already had the causes and the nature of that pain evaluated by a physician. If not, you may find yourself actually working outside of your scope of practice.

Pain Control Techniques

I like the term "pain control." It's the term Scott Sandland and Michael Ellner use rather than "pain management." Management just has a different flavor to it, a different interpretation by many people. Also, to doctors, "pain management" means prescribing medication for pain, so the term can lead to some misunderstandings.

Calling our work "pain control" lets your clients know that they can turn it up or turn it down. They can listen to it, hear the voice of that pain, and actually receive the message. Or they can turn it down to go take care of what they need to take care of. It tells the client they are going to have a level of control again.

So there are a number of different processes that I use with pain control. The first one I'll explain is really an NLP process; it's interactive and experiential.

A lot of hypnotists, for some reason, think that asking their clients to talk will somehow disturb the trance. It won't. Really, life is trance. We're always in trance. So the question a hypnotist ask is this: How can I utilize that trance?

So I have no problem talking to my clients during suggestive therapy and asking them to talk back. Hypnosis is not always me guiding them to an induction, guiding them to a deepener, and then reading a script to them. Sometimes hypnosis is an interactive process: sometimes with the eyes open, sometimes with the eyes closed.

If you are particularly worried that answering your questions will disturb your client's trance, you can always make this suggestion: "In a moment, you'll answer my questions in a clear voice, and the sound of your own voice will simply take you deeper."

I'm going to outline this process and then give you an example application of it.

Eight Step Pain Process

Step One: Get Commitment

Find out if your client is really committed to either change or ownership of the pain. Is your client committed to controlling the pain, to changing the quality of it?

Some clients find that there's a tremendous value in continuing their misery. It may bring sympathy. It may actually bring financial compensation. It may have a number of other benefits—psychological, emotional, spiritual—so the question is, are they ready to control their pain? When we have an affirmative then we can go on.

Step Two: Calibrate

Have your client identify what that pain is and then rate it: "On a scale of one to ten, with one being the lowest and ten being the most extreme pain you've experienced, how much pain are you experiencing right now?"

For the majority of my clients, it will be a six, seven, or eight. Now that might not be realistic by anyone else's measure, but it's important to accept the number the client gives you.

Step Three: Conceptualize the Pain

Ask your clients how they conceptualize, how they see the pain: "If your pain were an object, what would it be like? How would it look, feel, smell, taste, or sound? How big is it? What color is it?"

Now, I used the word "see," but really you want to involve all the senses and all the sensory submodalities. Get as much detail as you can.

Step Four: Identify the Purpose of the Pain

What's the positive intent of your client's pain? All pain has a purpose.

Step Five: Create an Alternative Representation of that Pain

Do that experientially. Ask your client to cup their hands and hold their hands out in front of them as they literally visualize the altered image or experience of the pain.

Step Six: Install the New Conceptualization

Have your client take those cupped hands and place the new conceptualization over that spot where they're experiencing the pain. Ask them if they have fully associated that new conceptualization into that place on their body.

Step Seven: Re-Calibrate and Break State

Ask your client to re-rate the pain on a scale of one to ten. Then have your client open their eyes for a moment (or close and open them) and really pay attention to that spot. Check to see how it feels different. Is that difference positive for your client?

Step Eight: Repeat if Necessary

I will usually have my client close their eyes and increase their comfort by going through the process again.

Example of the Eight-Step Pain Control Process

This is an abbreviated form of the process. The important thing is not that you remember or read it word-for-word, but that you absorb the interactive structure:

Close your eyes and relax in your chair. Simply breathe in and breathe out; relax the muscles in your body and bring yourself to that place of learning that you associate with hypnosis. Set aside any tension of the day, or worry about tomorrow, or concerns of the past. Really focus on your breath, simply breathing in, breathing out. As you continue to relax with each breath and each number, doubling that sensation of relaxation. Five, four, three, two, one.

Focus on a pain that you have identified as being difficult for you. The first thing that I'm going to ask is an important question: Are you truly ready to develop the ability to control, modulate, or even give up that pain forever? The question may seem silly at first, but really it's an important question. It's an important question because a part of us can often desire to hold on to these things for whatever reason. So simply nod your head if you're at that point, if you're truly ready to end your suffering by practicing the process that many others have found every useful. Great. Perfect.

Now identify that place in the body where you're carrying pain. Identify that spot and right now evaluate your experience with pain. On a scale of one to ten, at what level do you feel the pain? One being the lowest and ten being the highest you've ever felt in that spot. What are you experiencing right now?

As you continue to breathe in and breathe out, experiencing the pain at that level, hold your hands out in front of you, cup them together as if you're going to be holding something in those hands. And imagine the pain in that part of your body, that pain moving into those cupped hands. How do you see it? What color is it? Is it very heavy, hard to hold the hands up? Or is it very light? Is it small or is it large? Does it fill or does it not fill the hands? Is there a sound to it? Is there even a smell, a scent to the pain you're holding in your hand?

Because the mind is a remarkable machine, you have the ability to use the creative part of your mind to begin to change that pain. But before we change the quality of the pain, I want you ask yourself: Is there a positive message from this pain that I should be hearing? Is the pain telling you to take an action or giving you meaning? Or in some way or another, is it serving a purpose? If it is, and it doesn't always, but if it is, receive that message and be ready when our session is over to take action on that message.

Now as you hold that pain in your hands in the level, as you know its color, begin to see it change becoming either brighter, sharper, crisper, or clearer. Or possibly hazier, fuzzier, more difficult to note.

[What we're doing here is really changing the image so that our client conceptualizes it with the opposite with what they have been conceptualizing. So if they see it as small, make it large. If it's large, make it small. If it's making a noise, let it be silent. If it's silent, maybe play circus music. If there's a scent, perhaps a pungent scent or an unpleasant scent, let it become the scent of rose tea.]

And in every way possible, simply alter and change that pain so what you're holding in your hand is entirely different. It really is amazing how the mind has the ability to conceptualize these things. And as you have been holding this new pain, conceptualizing these things, I don't know if you've noticed that that place where you took it from has felt the sensation of no sensation or not. But take the new conceptualization of that pain—a different color, a different texture, a different flavor, a different set. Take those cupped hands or even one hand, if you need to, and simply set it on that place. Letting that new conceptualization seep into and become one with that place where you took it from. And now ask yourself this question: On a scale of one to ten, how does that pain feel? And note, of course, that the intensity of it is, if not completely gone, completely different.

Now go ahead and open your eyes.

[This is a way of breaking state while remaining in trance, a fractionation process. When I break state, I often simply have them open their eyes. It can be kind of jarring for some folks.]

So, open your eyes. Pay attention to me. Do you notice there's a difference? Yes. It's pretty amazing, isn't it? Yes. Awesome.

And go ahead and close your eyes again. With the eyes closed, allowing yourself to relax even further than what you're doing. You're doing great using that creative part of the mind. Five, four, three, two, one, zero.

From there, you can either repeat the process or transition into the rest of the session.

Radio Knobs

This is something that I've been using with my clients for twenty years, and they have found it tremendously useful. I used to actually work in an in-patient Pain Control Treatment Program. It was run by two psychiatrists in Houston, Texas, and I worked there part-time as a therapist. I used to work full time in the substance abuse program which was on the other side of wall, but I worked on a regular basis with them and I use this technique often. They found dramatic results even though it's a very simple visualization technique.

Close your eyes, and let yourself relax. Bring yourself to that resource state that we associate with hypnosis. Simply breathe in, breathe out. With each breath, with each number doubling that sensation of relaxation. Five, four, three, two, one, zero. Completely relaxed.

There was a time not too long ago when car radios didn't have remote controls, when we didn't have digital dials, when all of the radios actually had a knob on the left and a knob on the right. Because it wasn't too long ago, I'm sure that you have actually seen one at one point in your life, maybe even as a kid had a car like that with a knob on the left and a knob on the right. And the knob on the left of course controlled the volume level: 1, 2, 3, 4, 5, 6, 7, 8, 9, 10—the loudest level. The knob on the right controlled the frequency or the station that we were listening to. As you imagine those knobs, you can even see yourself turning the knob on the left and the knob on the right, where you can hear the changes as you adjust the volume and adjust the frequency. Maybe even you can feel the spinning of the chrome-colored plastic dials that seemed to have dominated the auto industry in a time gone by.

As you continue to breathe in and breathe out, think of your pain, the pain that brought you here today. You certainly have been listening to the volume of that pain over the last couple of days, and weeks, and months. Maybe even most of the time listening to it at level seven, or eight, or nine, or even ten. And so imagine that coming from the car stereo is the volume of that pain at the level where you normally hear it. And see yourself, taking the knob on the left and turning it down from a six to a seven, listening to it, feeling the vibration of the sound that it makes, and even seeing the number change from eight to seven to six to five.

Remarkable how the mind works, we can create an image of a car radio, an old-time car radio we can associate our pain into. And if you check your pain right now, you'll notice that if by turning down the volume a little bit further, you don't hear it quite as loudly. A six to a five to a four, maybe even all the way down to a two or a one, or maybe even turning it off completely, if even only for this exercise.

Now there's a knob on the right; in the car of course it changes the station from one station to another. We're going to use the knob on the right to move from pain to comfort. Because at all times there are both country-western and rock'n'roll on each and every radio; they're just in different places. There's always AM sports and always FM public radio on the same radio. But we choose what to listen to, what appeals to us. And I'm sure you've gotten in a car before where somebody else was listening to a station loud with something you didn't enjoy hearing. And so you turned down the volume and adjusted the frequency. Assuming your mind imagined taking that knob on the right and turning to the frequency of comfort. The frequency of comfort, perhaps you've listened to your pain so long you weren't aware that the frequency for comfort was present, but it is. And let yourself turned to that frequency feeling a sense of comfort. And now turn the knob on the left, increasing the volume of that comfort from a level one to a two, from a two to a three to a four to a five, even increasing the volume of comfort as high as you would like, to a level six or seven, eight or even ten.

Now I know in my office when you go through an exercise and experience the sense of change hearing now the volume of comfort no longer listening to the volume of pain, it may seem like a remarkable thing. But the mind is a remarkable thing. You see Napoleon Hill said that, "Whatever the mind can conceive, the body can achieve." And because of this reality, when our session is over today, when you step over the threshold of the door, when you get in the car, when you head back home or to the office, you'll find that you didn't leave the radio behind because it was a radio with knobs on the left and knobs on the right of your own creation. And so at any time over the next day or two or three, or even a week or two or three, or even over the next year or two or three. If at anytime, anywhere, you'd like to turn down the volume of pain and increase the volume of comfort, you can close your eyes for just a moment, bring yourself back to this resource state that you've created. Take control over your pain, but more importantly control over your comfort using these radio knobs anywhere and at anytime to find a sense of relief and control related to the pain that brought you here in the first place.

Glove Anesthesia

You may be familiar with Glove Anesthesia because you remember the 1970s psychology classes, where the professor would induce Glove Anesthesia—the sensation of no feeling in the hand—with a student. Then they would actually take a needle and pierce it through the hand, and the hand would bleed. Yet there would be no pain. It's a powerful demonstration.

Well, in our current litigious era, we no longer clamp our clients' skin, and we no longer pierce them with needles in order to prove to them that they're experiencing Glove Anesthesia. Really, that public demonstration was unnecessary, because the process of creating Glove Anesthesia is actually one that most of our clients have absolutely no difficulty producing profound response to.

Now what is the value of Glove Anesthesia? Truly, it can be used by hypnotists in a number of different ways. Of course it can be used as a convincer to work with a client, to help them understand the level of trance depth that can be achieved. This can help them to create resource states for intuitive healing and even metaphysical approaches to mind-body integration, and that's all fantastic.

Glove Anesthesia, though, can also be used as a concept for pain control. Our client, if they can create a sensation in the hand, can easily create a sensation anywhere in the body. And in fact it's even easier for them to simply transfer the sensation from the hand to the part of the where they would like to change sensation.

Here is a basic script for Glove Anesthesia. Again, I encourage you to experience it yourself before you use it with clients. Use the induction and deepeners that work best with your client, and then say:

Now being all of your attention to your hands. The back of your two hands—the left hand, the right hand; the right hand, the left hand. Your subconscious mind, even if the conscious mind sees them as the same, has already noted a difference between them. Perhaps it's noted the difference exists in the right hand, leaving only one hand left. Or perhaps you'll notice that change or difference in the left hand, and that will be the right choice for you. Either way is of course fine. But notice which hand is different than the other, and it is that hand that we're going to let lie still on the table or on your lap. Of course if you have not noticed a difference between either hand, you can simply pick the left hand or you can pick the right hand to lay on the table or your lap.

Now take your other hand and as you take your other hand, hold out your index finger as if you are going to point. And point to a spot on the back of that hand that you selected and just point to it. Just touch that spot at the back of your hand lightly. In fact, you can let your fingertip bounce up and down on that spot on the back of your hand—a spot a size of a fingertip, or perhaps the size of a dime. And you can tap it a few times. And as you tap it, notice the feeling of the tap. And now stop tapping, noting that very spot where you were pointing and the feeling are noting the sensation of no sensation in just that spot. It's amazing how we, in life, usually pay attention to pain or pleasure feelings, but rarely do we ever look for and note the sensation of no sensation—nether pain or pleasure, simply the absence of sensation. And notice that sensation in that spot that you were pointing to on the back of your hand, that spot about the size of a dime, and we can actually now increase that sensation of no sensation. Allow that spot to become larger, the size perhaps of a nickel on the back of your hand. And note that you're able to do that, even making it a bit larger the size of a quarter. Breathing in, breathing out. Noting the sensation of no sensation and a spot on the back of a hand the size of a quarter. Now expand it to a spot on the back of your hand the size of a silver dollar. Imagining the sensation of no sensation in that spot on the back of your hand the size of a silver dollar. And now expanding it to an even larger size, imagine that sensation of no sensation in your thumb, the fingers, the palm, even extending into the wrist. It's as if you have a glove covering your hand that has brought you the sensation of no sensation.

It's amazing that we can create this with our mind. This is of course not something that I've done to you, but something I've done with you, guiding you through. The value here is like Autogenic Training to see that we have the ability to create from within any experience at all. But what's awesome is that that hand that we have now noted the sensation of no sensation in from wrist to fingertip, from thumb to pinky, from front to back, can touch that part of us that has felt pain or discomfort, or is in need of healing.

Go ahead and lift that hand. Lift that hand, lift that hand, lift that hand. And simply place it on that part of you in need of relief or in need to healing, or in need of wellness. And you're going to let your hand rest again; noticing in your hand the feeling of feeling but in that place where you touched the sensation of no sensation, having been transferred to that place, that spot.

Chapter Five:
Sample Suggestions

I'd like to give you a few examples of how I apply these techniques with specific clients to meet their individual requests.

Hypertension

Hypertension, or high blood pressure, is often called the silent killer. It's something that causes many health problems, ranging from stroke to heart attack, to problems with impulsivity to sleep disturbances, to a number of different vascular diseases that are caused by untreated and often unidentified high blood pressure. If I have clients who are dealing with any medical issues, one of the things that I want them to do is to regularly take their blood pressure because it is very important that we monitor our blood pressure, especially when we start to go a little over, so that we can make sure that we can continue to be in optimum health.

When a person has high blood pressure, they will often present to us and say, "I'm here because I heard that hypnosis or meditation can be helpful with lowering blood pressure." And they'll say, "I don't want to take as much medication" or "I don't want to take any medication."

One of the reasons why men don't want medication is of course because of the sexual difficulties that can emerge as a result of both the short-term as well as certainly the long-term use of blood pressure medications. And women certainly have a variety of responses to these things as well. So there are a number of reasons why a person really wants to look to non-medication ways to control their blood pressure.

I want to stress here, I do not tell my clients, "OK, I'm going to teach you these techniques, so quit taking your blood pressure medication." This is the role of the physician. It is the role of the client to decide how they want to proceed. And so I want to advise my clients to continue ay medication as prescribed by their physician until their physician directs them to do otherwise. Remember as a hypnotist: scope of practice. I'm not practicing medicine; I'm not prescribing or treating or diagnosing. I'm simply providing training resources and ideas to my clients so they can embark on the path that is best for them.

So if my client said to me that his goal was to decrease his use of blood pressure medication, even eventually going off them, I would simply say, "That is really a great goal. A lot people share that goal, and the research shows that by learning the techniques of medical meditation and self-hypnosis, that by doing those things along with taking other steps to control your blood pressure, you probably will be able to achieve that goal." Or we can even be more direct and say, "You will be able to achieve that goal. I'm going to let you know that it is important for you to make those decisions with your physician. So I'm going to teach you the techniques, and the decision as to whether or not you should decrease your dose or get off the medication all together is one you should make with your physician. So let him know that you're learning these things and you're applying these things, and that you've been keeping a journal of your blood pressure so that he can work with you in making a decision, which is really what is best for you."

Again I'm not going to advise him to start, to stop or not. I'm not advising; I'm not prescribing. I'm simply training and teaching specific skills.

Now, I'm going to begin this sample session with metaphor and indirect suggestion. These are two suggestions that I'm abbreviating from a CD that Dave Parke produced three years ago called *Prairie Dogs*. And the reason why I'm using the elements from Dave Parke's script is I think Dave came up with two brilliant metaphors that are indirect suggestions for dealing with high blood pressure:

And so, as you continue to breathe in and breathe out, relax completely all the way down now. Five, four, three, two, one.

Before I give you some direct suggestions that can help you to achieve your goal of decreasing your blood pressure—notice I didn't say the goal of getting off your medication, that's for the client and the doctor to decide—*your goal of lowering your blood pressure, I want to share with you a story. In Kansas, there's an animal called a prairie dog. In fact many parts of the Midwest here in the US have prairie dogs. These prairie dogs are really pretty fun to watch. They dig tunnels that stretch literally for miles underneath the top of the earth. Some actually great cities underneath our great state have been built by these prairie dogs. And that's interesting because as a prairie dog burrows through the earth, he removes the blockages, the dirt, and the rocks, perhaps even chewing through the roots, leaving a perfectly clear tube underneath the surface of the earth where it can move.*

These tunnels can be quite long; in fact a prairie dog can actually run through those tunnels. And sometimes the tunnels connect to other tunnels. The prairie dogs reinforce the corners, and clear away anything that would make for an un-smooth transition. It's amazing how in a relatively short period of time, large amount of earth can be moved by these prairie dogs, clearing away any obstacle that's in their way.

Now, that's really a great metaphor for dealing with somebody with clogged arteries, neuropathy, other types of issues of problems that may have constriction in the vascular system. It is of course an indirect metaphorical suggestion that because prairie dogs are a part of nature and people are a part of nature, our bodies can of course clear away debris just as prairie dogs do.

The second part of Dave Parke's Prairie Dog CD actually talks about New York City. So I'm going to share that metaphor from Dave Parke's recording:

So breathe deep as you continue to relax. Five, four, three, two, one, zero.

Before I cover some things related to direct suggestion or the things what you can do to work towards your goal of lowering your blood pressure, I want to share with you an observation I made when I was in New York City. I was in New York City, and it was in the winter time and from my hotel I noticed that it was starting to become cloudy and it started to snow. My hotel actually overlooks Central Park, and it was really beautiful watching the snow fall. I think the snow must have fallen all night because when I woke early in the morning, the city seemed covered by a white blanket.

As I look outside of my hotel window, I can see that the city ploughs are busy clearing each and every street. They go up and down the Main Street, and they go in and out of the side streets and the alleys, paying attention to any area that was blocked, removing the snow early in the morning to make sure by rush hour that the taxis, the limos, the cars, and the motorcycles would of course be able to freely access all of those roads, highways and byways in New York City.

This is very similar to the prairie metaphor. It's a great metaphor for opening up, clearing out, removing blockages; a process to problem-solving. All of these are really good indirect suggestions.

After I do indirect suggestion with the client, I will almost always then do either experiential or direct suggestion.

So continue to relax. Five, four, three, two, one.

I'm going to share with you some, what we call direct suggestions; these really are things that I think you should do. These are actually things that you've told me you want to do.

I want my clients to take ownership of their direct suggestions.

These are things you've asked me to suggest to you by simply coming here today and being willing to address this issue. And we know of course that perhaps the best medication for blood pressure is sweat. And so by increasing your physical activity and even your exercise each day, you'll notice that your body responds to that by decreasing unsafe or distressing levels of blood pressure.

And of course if I've taught some techniques, this is where I try to tie direct suggestion to the induction.

I taught you some techniques earlier: Progressive Muscle Relaxation, Autogenic Training, and Mindfulness. Anytime over the next week or two or three you find yourself experiencing high level of either emotional stress or even physical stress, make the decision to take a moment to sit and still the body, relax the shoulders, even close the eyes. And return back to the state that you've created here today that we call hypnosis.

Direct suggestion number three:

Of course it's important to note your progress and so each day, at the same time each day and in the same position each day, continue to monitor your blood pressure, noticing because of the changes you've implemented and the strategies you've exercised the number respond each and everyday by decreasing any risk and any problems.

Direct suggestion number four:

And of course, because you care for yourself deeply, because you've committed to a valued path of making these changes, decreasing your caffeine consumption, replacing it with uncaffeinated tea or uncaffeinated coffee or uncaffeinated cola, and avoiding excessive salt and certainly tobacco, are all things you find easy to do, no longer of struggle because of how much you care for yourself.

Direct suggestion number five:

And each night, after you take your blood pressure, you can take a moment to practice the principles of Autogenic Training. Simply close the eyes and lie back on the bed, or even sit in a chair and let your hands be heavy, saying to yourself: "My hands are heavy. My hands are heavy. I'm letting those hands be heavy." And you can also say, "My hands are warm, my hands are warm, my hands are warm," and let those hands become warm. You can also say to yourself, "My forehead is cool, my forehead is cool, my forehead is cool," letting your forehead be cool.

In fact, take a minute to simply focus on your forehead and say to yourself, "my forehead is cool," noting the ability you have to sense coolness in your forehead. "My forehead is cool, my forehead is cool." In fact, you can even nod your head; did you note that sensation? Did you note that feeling of coolness in the forehead? Awesome.

And of course we know that if you can create warmth and heaviness, then you can create coolness. You can create clarity in every vessel of the body. You can create smooth and rhythmic breath and smooth and rhythmic blood. All of these are direct suggestions. I've given you, I guess, five direct suggestions here for dealing with blood pressure, and two metaphors or indirect suggestions that can be used.

And this is really the process that I go through with my clients to both teach them things that are useful to them. Notice when I do Direct Suggestion, I don't simply predict what they're going to do in the future:

"I want you to leave well and find that you exercise a lot. When you leave here well you'll find your stress level has decreased. When you leave here well you'll monitor your blood pressure. When you leave here well you'll decrease your caffeine, your tobacco, and your salt."

I don't just simply give the direction suggestions, predicting the future. I almost always tie those direct suggestions to either the present, other things that they've learned or to the values of the client. When I tie my direct suggestion to my client's core value, those direct suggestions are often acted on efficiently, quickly, and thoroughly by the client.

Cancer

Let me share with you five specific strategies for cancer. Let's assume that a man has come here because he's in the middle of his treatment for cancer. He's chosen this treatment, based on his own discussions with his doctor. I neither endorse nor oppose the choices clients make for radiation and chemotherapy and other types of treatments, including surgical processes. My role is to support their health within the context of the decisions that they made with their oncologists. This is how we remain within the scope of practice, not practicing medicine without a license, and really how we build referrals with clients. If an oncologist refers to you, and you tell your clients "with these techniques you won't need that radiation or chemo," you're not getting anymore referrals. That's a decision to be made between the oncologist and the client, not between the hypnotist and the client. My job is to support them no matter what choice it is that they make.

I'm going to share with you five specific visualizations and how I might use them with a cancer patient.

Here's visualization number one:

I don't know if you've ever been to the beach and tried to hold sand in your hand. But when you try to hold sand in the hand, the sand because of its water content is usually fairly weak. The more water it has, the weaker it becomes. Of course the opposite is true: if there's no water the sand becomes very weak and won't be held together either. It's almost as if there's a specific point where the sand and the water mixture is the strongest. Our bodies are like that. The healthy cells, the white blood cells, that are in our body those are the strongest. They're held together with the exact amount of the correct forces to sustain the life. But cancer cells are not healthy; they're weak. They're like that sand, one like you try to hold in your hand that has no water and so it simply falls in the cracks of the fingers. Or that sand that has too much water that simply drips through the fingers like soup.

Now focus on that place in the body where the doctor told you the cancer resides. And as you focus on that spot recognize that each tumor is composed of millions of cells, but these are unhealthy cells. They're not strong cells. They don't have the essential proportions of life-sustaining materials. So see them falling apart, unable to group together.

This is a very effective visualization. And it is one that can be used in a number of different ways. You can use sand on the beach for your analogy, or you can visualize those cancer cells as anything that is not strong and easily falls apart. You can compare cancer cells and healthy cells to a house of straw and a house of bricks. Then you can envision the wind of treatment blowing through, allowing the strong cells to remain strong and the weak cells to simply fall apart. These are all visualizations that could be used.

Visualization number two:

And as you continue to close your eyes, visualize the treatment you're receiving as powerful and strong. You've chosen to take certain medications and to have certain processes and procedures done, so however you visualize strength, imagine that strength coming into your body and doing what a strong and positive treatment can do—zapping or removing or eliminating those cancer cells, leaving behind exactly what we need to sustain life. In fact you could even visualize that treatment as a superhero.

Again an awesome visualization for your cancer patient.

Visualization number three:

Imagine the white blood cells being the good guys in white hats. And of course we know from watching childhood movies that the guys in the white hats always win. So, see those white blood cells as they are: the strong and good sustainers of life and see them in your mind attacking those cancer cells, always winning each battle and defeating the cancer cells in the black hats.

Visualization number four:

Each day during this course of treatment you've paid attention to the things that you need to do to take care of yourself. And let me congratulate you for making some serious changes in your diet and increasing your water consumption. And so you can actually visualize that any water that you're drinking is flushing through the body and flushing our cancer cells. Visualize the food that you're eating and the nutrition being dispersed through each cell of the body, the nutrition from that healthy food that you've been eating simply pushing out of the body, eliminating from the body any and all toxins, carcinogens, and even the cells of cancer themselves.

It's a great and simple visualization.

Visualization number five uses Future Pacing as a technique:

See yourself as you know you will be three months from now, six months from now, twelve months from now, cancer-free. You told me you when you came in here that you were going to do whatever it took to be cancer-free. Those were your words. "I will do whatever it takes to be cancer-free." And so, I know you have the ability to see yourself in the future free from cancer. As Napoleon Hill told us: "Whatever the mind can conceive, the body can achieve."

And so see yourself three months from now, six months from now, a year from now, receiving that message from your oncologist that you're cancer is gone. Many other people have received that same message, probably even from the same oncologist. And so as you look beyond the present and towards the future recognize that that "You of next year" is the same You as now. And so, associate into that step, into that letting that "You of the future" be the "You of right now."

All of these could be truly useful strategies for visualization with our clients who have the diagnosis of cancer and come to see us in clinical hypnotherapy.

Wound Care

Now let's talk about wound care. It could be a diabetic wound, a melanoma that's excised, a wart that's removed, or a puncture wound that was caused by an accident: The suggestions will be fairly similar in each case. And I want to share with you some ideas that can be used, both experiential indirect suggestion and direct suggestion, for dealing with this type of condition.

My house used to be next door to a dermatologist, so we would talk about these things every now and then. Having a melanoma excised can be a simple procedure—or it could be a difficult procedure for some patients depending on their health condition and other factors that come into play, the size of the area, and all these sorts of variables. So their patients can need our help.

Sometimes people say that hypnosis is a great tool for wart removal. Perhaps it is. But I can walk over next door to the dermatologist and have a wart zapped off in about five minutes. So I'm all in favor of recognizing that if there isn't a doctor around or I don't have insurance or I don't have access to the dermatologist, maybe using hypnosis for wart removal is a wise idea. But I would rather take five minutes, have them zap it off, and then use hypnosis for the healing process. You can make whatever choice you want to in life.

Here are some suggestions I would use for wound care:

I'm going to give you in a few minutes some direct suggestions that can help you use the skill of hypnosis in recovering from your wound. But first, I just can't help thinking about a great trip my family took to the shore.

And I had blast at the sea. I was there with the kids, and I was playing on the beach, and I really enjoyed it. Living here in Oklahoma we're not really near any water. It's always amazing to be able to watch the sea. And all day long the beach was full of people; they were playing in the sand, picnicking, building sand castles, driving up and down the beach. And the beach was really pretty rough at the end of the day. It looked like it had certainly been used.

From the balcony though where I was staying I could view the sea and I could see the high tide rolling in, the water coming in. And the tide would get a little bit hit higher, and a little bit higher, and a little bit more, a little bit more. And by the middle of the night the water was rolling almost all the way up to the road separating the hotel to the beach. And I noticed that each time the tide would come in and go out it would carry a little bit of the beach with it.

Then in the morning when I woke up I went for an early morning walk on the beach. When I took a early morning walk on the beach, I was amazed at how smooth it was, how the natural course of the ocean had simply cleared away the blemishes, the debris, the injury from the day's participants that had used that beach. It's amazing how nature has—in almost every aspect of our world—a unique ability to promote healing, health, and wellness.

Again that is an example of an indirect suggestion: a story, a metaphor, a parable that I might use with a client.

Now, I'm going to do something experiential using visualization again for wound care.

I told you the story about the sea. And as you continue to relax with each breath going into an even deeper state of hypnosis. With that creative part of your mind, bring your attention that place where the wound resides. With your eyes closed, bring your attention to that spot and see that wound as it is right now. Just think of the doctor, the incision was made; you know that the sutures remain. With that creative part of the mind and adaptive part of the mind, imagine what experiences you'll have after the period of healing and see that spot in the future healed, smooth, and safe. Are you able to do that? I'm going to ask you the question. Are you able to see the contrast in the images between the way it is now and the way you know it will be? Good.

Now go back to that image, that first awareness of your wound as it is right now. Certainly you've seen a time lapse video before. I was watching a video that had taken over the course of a year. It was amazing. It was a guy who took a picture each and everyday for a year and then condensed it down to a one-minute video. The time lapse of his hair changing, the time lapse of his beard growing. You've probably seen a time lapse video before, maybe a video where the sun rises quickly or the sun sets slowly. Have you see that type of video before, time lapse? Oh, I see.

I don't mind talking to my clients, asking questions. It's okay to ask them respond, nod, even answer verbally.

I want you to create a time lapse video or a time lapse image. See that wound as it is now and know how that wound will be when it is healed. And create a time lapse video really focusing on that spot where that wound is, seeing it on time lapse healing, healing, healing.

Now, some direct suggestions:

And so because you care for yourself, it will be easy for you to follow the doctor's instructions for caring for that wound. Finding that with each change of the dressing, with each washing, with each aspect of wound care that you engage in, the healing process is encouraged and you get better and better, faster and faster each and every day.

Direct suggestion number two:

And of course the body is comprised of many different macronutrients and micronutrients. Our bodies heal when protein levels are high, so the next couple of days, the next couple of weeks, you'll increase your protein consumption in your meals, making sure to take care of yourself from this perspective to promote that healing.

So we know that wounds respond not only to macronutrients but to micronutrients as well, like vitamins and minerals. So of course choosing foods that are highest in nutritional density is something that you'll want to do over the next couple of days, weeks or even over the next few years. Making sure you consume enough Vitamin C and choose foods high in zinc, you'll know that your wound heals faster, more comfortably, and with ease.

Those are all direct suggestions dealing with this specific presenting issue that we may be dealing with in medical hypnotherapy.

Global Neuropathy

Global Neuropathy is a condition in which a person has had damage of the nerves outside of the brain and outside of the spinal cord. This nerve damage can lead to number of different symptoms. The inability to sleep is one of the chief issues, and this relates to discomfort and other aspects of their condition.

Of course hypnosis is useful for sleep. People call me all the time and they say, "Is hypnosis useful for people with sleeping problems?" Of course it is. It's a great tool for sleep.

So when I have people who present with insomnia or global neuropathy and the inability to sleep or discomfort of any type, I actually teach them how to sleep. And this is one of the great uses of Progressive Muscle Relaxation. Some people say it's boring and it puts them to sleep, and say, "What's the purpose of that?" Well, sometimes clients need to go to sleep, so I teach people Progressive Muscle Relaxation.

One of the other presenting problems with global neuropathy is difficulty with balance because of the nerve damage, along with a constant pain or feelings of numbness. Now, earlier in the course, you learned Glove Anesthesia, which creates the sensation of no sensation in a spot. We can actually use the reverse process for the person with global neuropathy. We can have point them to that spot where they sensed the sensation of no sensation, and touch it, feel it, tap it, note it. If they can't sense it in the skin, can they note the weight and get some sense, some identification from a kinesthetic perspective to that. And then have them expand and increase the awareness of that sensorial experience. And so we actually do Glove Non-Anesthesia, teaching that strategy to a person and then having them put that where ever the numbness from the global neuropathy is.

So here are some direct suggestions that I might use for Global Neuropathy:

Over the next couple of weeks and the next couple of days, I know that you've been working out with a personal trainer, and so you'll find in the morning, when you get up a little earlier than you have in the past with the increased motivation for that strength training, saying to yourself: "I feel excited about getting well. I feel excited about getting well."

That direct suggestion is directly related to the fact that Global Neuropathy patients who engage in strength and conditioning training experience a greater state of wellness. If you have global neuropathy, strength conditioning is essential. So direct suggestions related to motivation for exercise or strength conditioning, more specifically, are more than appropriate with clients.

And so, when you get home tonight and your head hits the pillow that will be your indicator to your body that it's time to go sleep, and you'll find yourself drifting into a deep and natural sleep effortlessly tonight.

That's a direct suggestion. I'm telling him how he's going to respond when he has those cues.

EFT can be tremendously useful tool for those who have been given a diagnosis of Global Neuropathy. Here are some of the affirmations you can use:

Even though I have a diagnosis of global neuropathy, I deeply care for myself and appreciate myself.

Even though I have global neuropathy, I can experience wellness and happiness.

Or:

Even though I have a diagnosis of global neuropathy, I can experience wellness and happiness.

Autogenic Training is very helpful for Global Neuropathy:

As you continue to relax, you can think about how earlier during the induction I had you create a sense of warmth and heaviness and you were able to do that. You've been distressed by the feeling of numbness or tinglingness but just like you can create warmth and heaviness; you can create calm and comfort. With each breath, breathing into that place in the body where you felt numbness a sense of sensation, even saying to yourself: "I sense the sensation. I sense the sensation. I sense the sensation. I sense the sensation."

So we can actually utilize the components of Autogenic Training specifically to client. If they can create one set of feelings, they can create any sense of feelings. That's the amazing part of the human body.

Chapter Six:
Running a Medical Hypnotherapy Practice

How to Market Medical Hypnotherapy Services

Number One: Have a business card or flyer. I prefer flyers. People tend to save them and not throw them away or use them as a gum wrapper when they get to a restaurant, so you should have both. In every practice, the first thing you should do is order business cards so that people have a way to contact you and find your website.

Number Two: Have an excellent website. An effective website is a pre-requisite for good marketing now, but it's not enough merely to have a web-page; potential clients need to be able to find it.

A lot of my clients find me when they simply go to Google and search, "preparing for surgery" or "What is anesthesiology like?" You have to think about what kind of questions they have prior to surgery.

Then you need be sure that your website is optimized for the terms your potential clients search for; this is called Search Engine Optimization. Because of the way Google operates, local results show up first. Even if you live in a medium-sized or large community like I do, chances are that no one has optimized their web page for those questions. So it should be pretty easy for you to add a webpage to your website simply offering solutions for pre-surgical preparation and other terms that people search for.

Here is a quick SEO tip:

If your website is: www.AnyCityHypnosis.com

then you will want to create an extension to that, so it will be: www.AnyCityHypnosis.com/PreparingForSurgery

In other words, make sure the exact search term is in the domain name for that particular page. Then, use H1, H2, and H3 tags for those specific search terms. Next, write a unique, original article of about four hundred to five hundred words discussing that particular search term and add it to your webpage.

If all of this sounds daunting to you, you can hire a professional to do it for you, or you can teach yourself. In fact, the ICBCH actually has a video-on-demand series called *How to Build a Successful All-cash Private Practice*. In it, Video Number Two is devoted entirely to how I've created my webpage, how I go about Search Engine Optimization, and the type of tools that I use, such as WordPress, Google Analytics, and other resources.

So if these ideas that I'm mentioning right now are actually going over your head, what that means is that you need some additional education in the area of Search Engine Optimization and selecting the proper tools to build a website, which of course is not the purpose of this course. You'll find lots of resources on HypnoThoughts.com.

Almost all of my webpages are built with WordPress. And the reason I like WordPress is it gives me ability to add those four hundred to five hundred word unique articles that reflect that specific search term and provide information.

What Google is looking for is not a brochure or online advertisement; they're looking for content. So for your website to be brought to the top what you need is valuable content, how-to articles; that's what people are looking for. And of course on that webpage where you have that article with that information related to that search term, you'll want links to your main page, the services you provide, your telephone number for further information, and an email contact. I get a tremendous number of referrals from those who simply click those articles on my website and send me an email inquiring further, asking for further information, usually supplying both their email address and their phone number.

So, basic Search Engine Optimization is an essential strategy. And if you don't already have those skills, they're not particularly difficult to master. You can find the resources whether they're from us or others to help you develop your potential in that area.

To give you an example of how I use this to market medical hypnotherapy, on my website, one of the things that I offer is hypnosis for pre-surgical preparation. So I have a page with a short article that points outs some specific facts, including that those who learn the techniques of medical meditation or self-hypnosis prior to surgery have three predictable outcomes: decreased complications, faster recovery, and less dependence on medication. This is really important as we grapple with the issue of pain pill addiction post-surgery for some types of surgical clients. So people come to my website looking for this information, and they find that I offer a specific service.

Appendix B is an list I created called *44 Ways to Market Your Hypnosis Practice*; it is a pretty good list of list of marketing ideas, many specifically related to medical hypnotherapy. Here are some of the highlights.

Advertise in your local newspaper, especially if you live in a small area where large ads are very affordable in print media. By the way, always negotiate with the newspapers. Newspapers are hurting right now. They're willing to give you a bigger space for less money. Never take their first offer. I have advertised in the past in the classifieds section and in those weekly classified newspapers. That type of advertising is a very inexpensive. You can afford to advertise usually sixteen weeks or thirty-two weeks at a time. It becomes like a resource or directory; people always know it's there. So when they have a need, they say, "Hey, didn't that hypnotist advertise pain control?" or "Didn't he advertise pre-surgical preparation? I'm going to go get that classified ad and look for it."

Approach cancer centers as an adjunct staff member or as a person within their Patient Education Program. Almost every regional and local cancer center has a Patient Education Program. The National Institute of Health and most organizations that provide information, resources, and services on cancer treatment all advocate teaching meditation and self-hypnosis to clients. So if you approach a cancer clinic, with a well-worded cover letter, a professionally produced brochure, and a professional appearance, to offer your services, either as a give-back that you do for the community or potentially even a paid position, they are likely to be receptive to you.

Approaching pain clinics is also an idea that is useful. I have a friend who owns several medical clinics. He's got a Master of Business Administration; he's not a medical person. He hires doctors, he own clinics, and he owns a pain control program. And in order to provide the medications that the patients are often seeking at that program, he must, in order to comply with state law, also provide additional resources related to pain control for the client. That is the type of person who would certainly be open to using a hypnotist to provide that types of services.

Speaking to local civic organizations is also an excellent idea. Leave your brochures, your business cards, your information, and even information about seminars that you might be doing for patients or support groups that you might start on local health food bulletin boards.

Newspapers are always looking for new articles, and not too much happens here in Tulsa, so the newspapers and the TV stations, especially the six a.m. news, they're looking for information they can pass along to others. Ask to speak to a lifestyle or the health editor of the newspaper.

Call the TV stations, and ask to speak with the morning show producer. At least get their email address, and let them know what you do and that you would be available for interview. It's very effective.

How to Structure a Medical Hypnotherapy Program

When clients come to me, I want them to commit to a course of success; by pre-paying for a certain number of sessions, they commit to that course of success. Now one of the things that I do not do is make people commit to a long service of hypnosis. In fact, I think hypnosis is valuable because traditionally it is a brief intervention. I rarely ever spend more than three to six sessions with any of my clients.

So I offer a three-session package, essentially a training program, to my medical surgical clients so that they can, prior to surgery, master the skills that the research shows are effective tools for recovering faster, decreasing complications, and relying on medication to a lesser extent.

People are dreading surgery because it is going to mean down time. They want to recover quickly, they want to avoid complications, and they're looking for resources to help them. When they find those citations from the journals showing that the techniques I teach are valuable techniques, they call me up often without doctor referrals and say, "Hey, I'd like to do this. I'm having surgery in a couple of weeks." And so a lot of my medical clients are actually self-referrals rather than clients who are referred to me by medical providers, although I certainly do have those referrals as well.

I currently charge $359.00. How much you charge is going to be dependent on your geography more than anything else. How much you charge really probably is a function of where you live. If you live in New York, charge $750.00. If you live in LA, charge $500. If you live in Tulsa, Oklahoma, charge $359.00. If you live in Benkelman, Nebraska, charge $129.00. Really, money doesn't translate geographically. That's something I'm always stressing to people when they ask me how much should I charge. A million dollar house in New York City is a lot smaller than a $190,000 house that I have here in Tulsa, Oklahoma. So, money is really a function of geography rather than anything else.

I charge $359.00 for those three sessions. And in the first session I teach Mindfulness. And I do hypnosis sessions with them teaching them self-hypnosis, giving them a couple of CDs or MP3s to reinforce that learning. In the second session what I do with them is usually spend the majority of my session teaching them Autogenic Training.

In the third session, I'll usually teach additional self-hypnosis and meditation techniques. I might teach Object Meditation or Mantra Meditation, where we seek our attention focus on either an object or a specific word. And the research of course shows that this type of meditation is a very valuable skill.

Now, people sometimes say to me: "What's difference between meditation, hypnosis, and yoga?" And I've thought about this long and hard, and the only answer I can come up with is in yoga we wear funny clothes. I really think that meditation, yoga, hypnosis all might look a little bit different, but the reality is they all are essentially the same process.

So I teach a combination of meditation and self-hypnosis techniques to my clients. Now I also will spend time with them giving them both indirect and direct suggestions during a formal trance process to address the specific issues that they have come with. But really that three-session package that I work with is educational in nature.

Now, there's nothing that's written in stone. All of these things can be adapted. In fact somebody said to me, "You talked about Autogenic Training. But I have menopausal women and when I ask them to their warm their hands, they would rather get rid of those feeling." Well, don't use that technique with somebody who finds that technique uncomfortable. Nothing is absolute. Nothing is written in stone. We can always adapt to the unique needs of our particular clients. I do that not only clinically with my suggestions and techniques that I offer, but I also do this in the structure of my program.

Not too long ago, I was referred a woman who was having major surgery related to her reproductive system. She heard about my services three days prior to her surgery. Obviously there wasn't enough time to do three sessions with her prior to her surgery, so I spent about two hours with her in that first session about two or three days before surgery addressing her pre-surgical anxiety, giving her some direct suggestions and some skills that she could utilize.

I taught her some specific techniques that she could use following surgery, related to, again, Mindfulness, Autogenic Training, and Visualization. And I gave her a couple of MP3s and a 90-minute DVD on Autogenic Training; I gave her that DVD because I figured after surgery she wouldn't be going anywhere, so she actually had time to watch the teaching videos.

So I gave her that video, I gave her a copy of my book *Medical Meditation*, which I wrote a few years back. And I said to her, "After you return home, there's obviously going to be a period where you're not going to be mobile and you're not going to want to get out. If you want to, for the second session during that time period, I would be more than happy to meet with you via Skype. And if not, after a few weeks of recovery using the things that we've talked about today and using the resources I've given you, we could follow up here in my office as soon as you're ready to get out." And so I did that second session and the third session in week three and week four following the surgery she just had, and she found it helpful of course.

So you could adapt this to your clients. You could do a single session; you could do eight sessions. For example, I often talk about Mindfulness Meditation and Medical Meditation. They use those words interchangeably. Jon Kabat-Zinn has an eight-week program for teaching mindfulness-based stress reduction to medical patients at the University of Massachusetts. I like threes and sixes, so I generally work with a client for six weeks, but Kabat-Zinn works with his clients for eight weeks. I'm sure somebody out there has added their own spin and does it in twelve weeks. All of these things can be adapted to our unique clients, our unique personalities, the type of services that we offer clients, and their level of severity. There's a big difference between working with somebody who is really wrestling with a chronic life condition that they've given up hope on, and a patient who is facing, although major surgery, a process that is going to correct an issue so they no longer have a problem following that surgery. These are two different experiences. These are two different attitudes. These are two very different clients.

There is no protocol. There is no manual that says, "Buy this manual for $59.95, read this script or do this over two weeks, or six weeks, or eight weeks with your clients and they will all magically get well." In fact if you ever see a program promising that, run as fast as you can because it's probably not a program that's truly beneficial to your clients. All of these things can be and should be adapted to their individual situations, their individual needs, as well as your skill level.

As we talk about therapist skill level, I think it's important that you realize the value of mindfulness meditation. In addition to the exercise in this course, you can learn more at mindfulnessmeditation.org.

You want to integrate this with your clients, even if your skill level in adapting mindfulness is at a basic level. That's a great starting point. You can include this as part of a session rather than a full session, or certainly rather than an eight-week protocol. It's perfectly okay to function within your skill level, mixing techniques and ideas that can help our clients.

Appendix A:
The Nongard Assessment of
Primary Representational Systems

1. When you are injured, what is your immediate response?
   a. See the world as if it is magnified
   b. Hear the sound of impact
   c. Feel the sensation of pain

2. When you spell a new or difficult word, do you:
   a. Visualize it on a blackboard
   b. Sound it out
   c. Start writing it out

3. When you read, do you:
   a. See images of what you are reading
   b. Have conversations with the characters
   c. Seek stories with action and behavior

4. When you think, do you:
   a. Imagine your thoughts as a movie
   b. Hear yourself talking to yourself
   c. Become distracted by external activity

5. When driving, do you:
   a. Daydream in pictures
   b. Listen to talk radio
   c. Rock out and dance

6. If you buy an assemble-it-yourself project, what do you do:
   a. Look at the picture on the box
   b. Read the directions out loud
   c. Just start building and complete it by trial and error

7. When you go to movies or watch TV, do you:
    a. Prefer rich scenery of distant places
    b. Enjoy the dialog of heavy movies like court dramas
    c. Get bored and wish you could go do something else

8. What produces a more intense sexual response?
    a. Erotic images
    b. The sounds of lovemaking
    c. The feeling of skin touching skin

9. When you give a speech, do you:
    a. Talk with your hands
    b. Hear yourself telling you what to say
    c. Speak slower than other people

10. When relating to others, do you:
    a. Imagine them taller, fatter, further, closer, or different in any way; or pay particular attention to unusual features they possess
    b. Find it easy to follow the stories, jokes and conversations with others without feeling lost
    c. Move toward them, feeling their energy

Appendix B:
44 Ways to Market Your Hypnosis Practice

by Richard K. Nongard, LMFT
(reprinted with permission)

What follows is a list of 44 proven strategies to reach the Medical Hypnotherapy market.

1. Buy radio ads, and negotiate the rates. They are cheap IF you are a good negotiator.

2. Buy classified ads and run them week after week.

3. Create a niche. Here is a 90 minute lecture on how to do this. http://www.hypnothoughts.com/group/icbch/forum/topics/how-to-find-hypnosis-clients

4. Get a folding sandwich sign for the road by your office. Put it up each day.

5. Get a business card with your picture. People respond better to them.

6. Create a good webpage, with a geographic domain.

7. Learn the keywords people search for, use them on your webpage titles.

8. Ask your personal physician for referrals. You already have a relationship.

9. Join a gym, become a model for your clients. You will meet people there trying to make a change, give them your card.

10. Make a really nice brochure. Here is a link to a template http://www.hypnothoughts.com/group/icbch/forum/topics/how-to-create-a-brochure-for

11. Buy an ad in small local papers selling ONE service.

12. Ask friends to recommend your service to others.

13. Create a topical group on Facebook and offer advice and information.

14. Start offering services online so you can get clients worldwide.

15. Join a local small business network and actually go to meetings.

16. Join the chamber of commerce and actually go to meetings.

17. Go to all the offices near yours and give them a couple flyers and brochures and introduce yourself as their "neighbor."

18. Contact groupon.com or livingsocial.com and run a promotion.

19. Rent a billboard. Never put more than 9 words and a phone number.

20. Use meetup.com and create a weekly group.

21. Join other groups in your area related to hypnosis, and become a patron of those businesses. Like the nearby yoga studio for example.

22. Write local press releases about your service.

23. Contact the local TV station lifestyle news editor. Tell them you are free for an interview.

24. Put a professional sign or wrap your car.

25. Create a newsletter.

26. Ask old clients for referrals, it's easier to get clients from an existing pool of clients.

27. Give away free CDs that have your contact on it.

28. List yourself in Google and Bing "places."

29. Call local community organizations, volunteer to speak.

30. Volunteer or join a board for local community organizations.

31. Invest in a professional office and furnishings.

32. Buy books and read them, related to consulting and business development.

33. Make a blog with useful info. Keyword it to your geography.

34. Create coupons and give them to your hairdresser (tip well).

35. Tip waiters well, and leave your card.

36. Do a postcard bulk mailing for the addresses near your business.

37. Use advertising packs that come in the mail with other local businesses.

38. Sponsor school activities and get your sign on the court or field.

39. Sponsor local sports teams.

40. Cross promote with other businesses. You market my clients, I'll market yours.

41. Buy bus bench ads.

42. Become a member of a church or other house of worship and teach a class.

43. Cold call other business professionals who might refer to you – DO IT!

44. You do business at other businesses. Tell them what it is you do, and tell them to tell the employees what you do.

Bibliography and Reading List

Lipton, Bruce H. The Biology of Belief: Unleashing the Power of Consciousness, Matter & Miracles. Carlsbad, CA: Hay House, 2008.

Nongard, Richard K. Medical Meditation: How to Reduce Pain, Decrease Complications and Recover Faster from Surgery, Disease and Illness. Tulsa, OK: Peachtree Professional Education, Inc. 2010.

Parke, David Prarie Dogs. Audio CD, 2011 www.LifeAfterFear.com

Rossi, Earnest L. Psychobiology of Mind-Body Healing: New Concepts of Therapeutic Hypnosis.

New York, NY: W.W. Norton & Company, Revised Edition, 1993.

www.ingramcontent.com/pod-product-compliance
Lightning Source LLC
Chambersburg PA
CBHW022108170526
45157CB00004B/1530